Multi-Hazard Disaster in Japan

PETER LANG
PROMPT

PETER LANG
New York • Bern • Berlin
Brussels • Vienna • Oxford • Warsaw

Dylan Scudder

Multi-Hazard Disaster in Japan

Part of the Pentalemma Series on Managing Global Dilemmas

PETER LANG
New York • Bern • Berlin
Brussels • Vienna • Oxford • Warsaw

Library of Congress Cataloging-in-Publication Data

Names: Scudder, Dylan, author.
Title: Multi-hazard disaster in Japan: part of the Pentalemma series
on managing global dilemmas / Dylan Scudder.
Description: New York: Peter Lang, 2020.
Includes bibliographical references.
Identifiers: LCCN 2019041567 | ISBN 978-1-4331-7530-5 (hardback: alk. paper)
ISBN 978-1-4331-7558-9 (ebook pdf)
ISBN 978-1-4331-7559-6 (epub) | ISBN 978-1-4331-7560-2 (mobi)
Subjects: LCSH: Social responsibility of business—Japan—Case studies. |
Crisis management—Japan—Case studies. | Emergency
management—Japan—Case studies.
Classification: LCC HD60.5.J3 S38 | DDC 658.4/08—dc23
LC record available at https://lccn.loc.gov/2019041567
DOI 10.3726/b16466

Bibliographic information published by **Die Deutsche Nationalbibliothek**.
Die Deutsche Nationalbibliothek lists this publication in the "Deutsche
Nationalbibliografie"; detailed bibliographic data are available
on the Internet at http://dnb.d-nb.de/.

This publication has been peer reviewed and meets
the highest quality standards for content and production.

29 Broadway, 18th floor, New York, NY 10006
www.peterlang.com

Every action has its pleasures and its price.

—Socrates

Contents

Figures

Preface

Multi-Hazard Disaster in Japan provides readers with a hands-on approach to dealing with the complexities unique to international management. The book draws strategically on this complexity to help readers cultivate greater critical thinking. It goes beyond the usual static business decisions related to setting the right price or memorizing truisms of management, and chooses instead dynamic real-world cases that require serious deliberation and creative problem-solving. While the book has been fictionalized, it is based on firsthand fieldwork and rigorous research on actual business cases.

Each of the three chapters starts out with a brief background of the context and the company, and then confronts the reader with a formidable and escalating challenge to work through. The scenarios include a high degree of complexity that companies often face in fragile or uncertain environments, whether in developing or industrialized countries.

With a focus on cross-sector collaboration, and managing the tensions between financial, social and environmental priorities, the trilogy helps readers learn how to develop creative solutions to seemingly insurmountable dilemmas. The book contributes in this way to management skills as well as to the knowledge needed to resolve issues in economic development, peace and conflict studies, political science, international relations and related fields.

Although the topics deal with high levels of complexity, they are written in an engaging and accessible way for both current and future managers interested in stakeholder management, business in developing countries, Corporate Social Responsibility (CSR), sustainable business, international entrepreneurship, strategic leadership, global business issues and related subjects.

Multi-Hazard Disaster in Japan is written as a trilogy to let readers experience how a business challenge evolves over three distinct phases. This design enables readers to "stack" their learning from one part of the book to another. As business issues take on an increasingly global character, the material is organized to help the next generation of business leaders prepare for a wide range of scenarios in uniquely challenging environments.

Disclaimer: This book is intended solely as discussion material. None of the content is intended to serve as examples of effectiveness or ineffectiveness of any kind. Though inspired by real-world events, many of the names of the people, groups and other entities appearing in this work have been changed to preserve anonymity. Any resemblance to real persons, living or dead, is purely coincidental.

Acknowledgments

It was Johan Galtung, "Johan" as he preferred, who introduced me to conflict theory as a way of applying creative thinking to seemingly impossible problems, and as a way of making work and life more meaningful. Johan recommended me to an MA program in peace and conflict studies in 1996, which I enrolled in the following year. For the next four years, I traveled extensively with him to mediation interventions, lectures and workshops in Okinawa and across Europe. Johan offered me a position in his NGO, Transcend, and advised me on my master's thesis that explored the Transcend method of conflict transformation. During this time, he also arranged for me to do an internship at the United Nations (UN) in Geneva, which opened for me a rewarding career path at the UN and later in business and academia. Those first years with Johan were some of the most intense, educational and adventurous ones of my life. Each day was filled with hard work, friendship and more humor than two people should ever be allowed to enjoy.

Ian Green was Chief of the United Nations Training Section in Geneva when he hired me as a training coordinator. Ian generously gave me the space to experiment and discover how to create productive learning environments. Though this proved quite costly for him at times, he always went out of his way to make sure I had everything I needed to grow professionally. I can do the work I do today only because of the opportunities I was given in that first stage of my career.

I want to thank Jane Nelson, both for the encouragement she provided me through her book, The Business of Peace, and especially for the personal guidance she gave me. Jane invited me to her office at Harvard University, where she shared with me inspiring insights into the field of business in areas at risk. Much of what I learned from her writings has illuminated for me how ethical decision-making impacts on firm value.

Kiyoshi Amemiya, the inventor of the now-famous demining equipment that transforms minefields into orchards, gave me a once-in-a-lifetime chance to work with him and his team of engineers on a mine-clearing project in Colombia. He provided me with an insider's perspective—even letting me test-drive the mine-clearing equipment to better understand some of the operator's challenges first-hand. The onsite experience helped me appreciate the complexity of working for peace in a way that is also financially sustainable.

In the very volatile days following the Great East Japan Earthquake, it was Ray Bremner OBE, then CEO of Unilever Japan, who offered me a seat at the table where he and other members of the crisis response team would meet to review and plan. This life-changing experience gave me a chance to see some of the leadership behind Unilever's promise of Doing Well by Doing Good in the extreme conditions of disaster relief. This book is very much an attempt to let the reader sit at that same table and join in the decision-making.

Finally, I owe a particular debt of gratitude to my editor and friend, David Hulme. David was helpfully relentless in getting me to think and write more clearly. Visiting David and Satoko for hikes and in-depth discussions at their home in the mountains on the outskirts of Tokyo provided me with the extra push I needed at critical moments along the way.

DYLAN SCUDDER
Tokyo
August 2019

Introduction

Multi-Hazard Disaster in Japan aims to help business leaders of today and tomorrow learn how to stay profitable in an increasingly volatile world. Drawing on the latest risk management research and direct participation in disaster relief in the Great East Japan Earthquake of 2011, this book provides an immersive reading experience based on the real-world case of a global company responding to the highly uncertain environment that emerged in the wake of that disaster. The central challenge is reconciling the competing demands of diverse stakeholders: immediate and long-term profitability, social concerns and environmental priorities. The book is grounded in the insight that such issues are always interdependent. As such, companies that ignore their stakeholders put their future viability at risk.

This is easy to say, but much more difficult to realize. Following the PDCA model made popular by W. Edward Deming and refined in Japan, the book is organized around the four phases of *Planning, Doing, Checking* and finally, *Acting*. Within this framework, readers develop competencies by working through business challenges that they prepare for, engage in, reflect on and then draw lessons from that can be applied to the next challenges they encounter. Readers work as managers at Pentalemma Inc., a major risk advisory which has a number of large multinationals in its client base. This book is part of a larger collection of books, each covering a different location and linked to a specific industry facing its own unique set of challenges. Each book is written as a trilogy with three distinct phases within each book.

The main question for readers is how to select an optimum business strategy from among several that emerge over the course of the session. They receive a series of reports and press releases, which generate a realistic working environment and expose them to modern business tools. Parallel to this, they are informed by various advisors, such as business mentors or problem-solving specialists. The experience of reading the book is similar to the experience of managing other urgent problems at work, namely, through a flurry of emails and discussions. Immediately after preparing for the upcoming scenario, it's down to business. Rather than the traditional outdoor adventure, classroom lecture or hi-tech animation, readers are embedded in a corporate team and focused on a complex issue that must be solved without delay.

Learning is optimized by immersing readers in high-impact business challenges through intense but entirely plausible scenarios. These might be set anywhere in the world—from Colombia to Japan. What they have in common is the struggle of making critical decisions in the face of uncertainty. Readers must find practical solutions within quickly changing, high-risk situations. By the end of each chapter, readers will have chosen the plan they think their client company should adopt. The outcome, in the form of client feedback, company results and broader social reaction, will be shared with them through review material that describes plausible outcomes to the decisions they made. Finally, they are provided with several pages in the back of the book where they can make notes of their progress and create an action plan for further development.

In short, the book emphasizes risk awareness and mitigation by helping current and future managers understand that failing to consider the potential consequences of their everyday decisions exposes them and their stakeholders to unnecessary risk.

Chapter One

Disaster Management

You are a manager at a company called Pentalemma Inc., which specializes in stakeholder management. The company took its name from the Greek *penta* (five) and *lemma* (assumption), the forgotten wisdom of identifying five solutions embedded within a problem and then choosing the best one.[1] Pentalemma's headquarters are near Geneva, Switzerland, with several hundred employees around the world. The most frequent problem that Pentalemma's clients have is that their stakeholders disagree with each other. Pentalemma's competitors are often unable to help their clients find solutions because they can only see two solutions. For them, all stakeholder problems are "dilemmas."

Pentalemma managers offer a better service because they can see five solutions to stakeholder problems. They find the client's most difficult problem, usually a disagreement between their stakeholders, and then create five options for them to choose from. Clients try to choose the best "win-win" option for themselves and their stakeholders.

As a Pentalemma manager, you help your clients improve their business performance by working with them to resolve as many stakeholder dilemmas as possible. Most of Pentalemma's clients are big companies with offices in many countries around the world. Some of these offices are in unstable or dangerous places. Your current client is Brighter Brand Japan (BBJ), a subsidiary of Brighter Brand Group (BB), a large multinational with a wide spread of interests and activities across the country.

Basic Background on Brighter Brand

Company History

Based in Birmingham, UK, Brighter Brand (BB) is the third-largest consumer goods producer in the world. The company began in 1888, when Donald McFarland established a food manufacturing concern in Birmingham.

McFarland had originally trained as a chemist and had become interested by the new opportunities that were opening up in two distinct fields: industrialized food processing and food preservation. His range of canned meats was soon being sold to customers from across the globe.

From canned meat, McFarland Foods soon expanded into producing chocolate and confectionary, all from its one plant in Birmingham. But the biggest change during that early era came with the merger in 1893 with Carlton Cleaning Products. Carlton, also based in Birmingham, produced soaps and chemical cleaning agents. McFarland partnered with the firm to produce the first mass-produced perfumed bath soap. Brighter Brand was born.

By 1914, BB was a public company quoted on the London stock exchange. It had plants in Birmingham, Manchester and Sheffield. World War I, with its huge demand for manufactured consumables, was to see the company double in size. As new markets and new technologies opened up during the middle of the 20th century, BB expanded both its range of brands and its production and distribution facilities. An American subsidiary, BBUS, was launched in 1935. BB managed to buck the trend of the great depression and continue to expand, relying on value and convenience to penetrate new markets.

After World War II, the company developed into a truly global enterprise. It continued with the federal model established with BBUS and it now operates through 15 regional and national subsidiaries, including such names as BB Japan and BB India. Taken together, the global company now produces over 150 well-known brands. It now employs 170,000 people all over the world and had a turnover last year of $52 billion. BB has four product lines; construction equipment, foods, cosmetics and cleaning products. BB also invests heavily in research and development.

BB sees itself as an innovative manufacturer and distributor of consumable goods. Harking back to Donald McFarland's original business model, BB's job is to take raw materials and combine them into products that people will use and buy again. This implies a significant commitment to Research and Design, especially given the fragmenting and increasingly sophisticated global consumer market. McFarland's eye for the potential offered by new technologies and raw materials continues to shape the company's development outlook. *Bellisimo* instant pasta and the *Just You* beauty range are two recent examples.

The nature of the business also means that BB is one of the world's largest purchasers of commodities such as coffee, tea and palm oil. By the same token, it is in its own right a major shipping and haulage company.

BB's product range divides into four sectors. These are foods, confectionery, cosmetics and cleaning products. Production is distributed globally according to demand and supply, with the subsidiary companies taking lead status for those products allotted to their area. BBJ, for example, produces the *SuperWok* range of Asian foods for the entire global operation.

Mission and Values

BB describes its mission as "Improving people's lives by providing consumable goods which offer quality, utility and value."

This mission is delivered through the company's values:

> "Innovation and Enterprise—We believe that our world never stands still; and nor can our business.

Fairness to all—All of the brands taken to market by BB offer good value. But in addition, we attempt to improve the world that we all share and to treat customers, staff and people everywhere with fairness and respect.

Long-term value—BB will ensure long term value for its investors, those who ultimately own the company. A glance at BB's dividend and share value history will evidence this commitment."

BB often reminds the public that all its business activities must be good for people, good for profit and good for the planet.

Governance

BB is managed by a Global Board (GB), which generally meets in Birmingham. The GB has 12 members, including four non-executives. The current non-executive Chairman is Jocasta Richmond. In turn, each of BB's 15 subsidiary companies has its own Board, of similar makeup. These are based in the regions concerned, such as BBJ in Tokyo and BBUS in Chicago.

The company's global accounts are registered at Company's House in London on 31st March each year. BB's auditors are Dixon and Lockwood Plc. The BB Annual Meeting takes place in March, in London. BB has a Global Audit and Scrutiny Committee, chaired by non-executive Board member Dilip Patel. BB has a policy of making as many of its records available to the public as possible. These are published online at the company's website. The Company's accounts, annual reports and its "Staff, Society and Planet Strategy" are all public documents.

The Brighter Brand Deal: Corporate Social Responsibility

Although not a Quaker himself, Donald McFarland was deeply impressed by the Quaker business model which he saw for himself in 19th century Birmingham. Anxious to avoid association with a specific religion, McFarland nonetheless insisted that the overall welfare of his staff and their families was a company responsibility. Workers' housing was built, alongside facilities such as a factory theatre and social club. Health services were provided. In various ways, this attitude still imbues the company ethos to this day.

BB has moved forward, however. In two important respects, the modern multinational has broadened McFarland's vision. First, BB is interested in the overall welfare of the communities in which it operates. It seeks not only to avoid doing harm but also to put something back—to make a direct investment in those societies with the greatest need.

Furthermore, since 1989, BB has added another dimension to its activities: the protection and enhancement of the environment. Protection in the sense that the company aims to ensure that ultimately all of its extraction activities are sustainable. An example of this was BB's commitment to secure 100% sustainable tea production by 2018. An example of enhancement is the Better Brand Chimpanzee Sanctuary in Ghana.

The Brighter Brand Deal draws these three strands of "Corporate Social Responsibility" (CSR) together into a single strategy: Staff, Society and Planet (SSP). All BB business planning is specifically required to actively incorporate the SSP strategy. Although this approach can bring greater short-term cost, BB recognizes this and, importantly, believes that it nonetheless makes sound long-term business sense. The strategy is endorsed by the Global Board and reviewed by it annually.

Trends in Planning and Development

In considering its medium- and long-term future, BB's Global Board has identified three priorities. These represent the basis for business planning, subject of course to changing circumstances. They also signal the overall strategic posture of the organization, intended to provide assurance to communities, investors and staff:

Already well-established as a mature player in consumer goods, BB intends to maintain and enhance its position. Consolidation and reflection, as opposed to a dash for high-risk growth, will characterize much of its planning. As the consumer market becomes more fragmented, BB will attempt to meet a more sophisticated pattern of demand. Niche products, such as those designed to appeal to the growing Islamic market, will form an important part of this strategy.

BB continues efforts to configure itself as a rather different multinational. In an era in which large companies are (with some justification) regarded with suspicion and hostility, BB intends to remain in the vanguard of sustainable and ethical capitalism. It will ensure that it pays appropriate corporate taxes in those countries in which it works, for example. This theme will be part of the Brighter Brand Deal.

Brighter Brand in numbers	
Full-time equivalent staff employed across the globe	
at 1st January this year:	170k
R&D expenditure, last year to this year	$1b
Turnover last year to this year	$52b
Growth in turnover:	
Three years back:	3%
Two years back:	2.8%
Last year:	3.7%
Share price at 31st January this year	$16.31
Share price at 31st January last year	$15.80
Share price at 31st January ten years ago	$12.14
Ordinary share dividend from last year to this year:	$1.37
Company registered at: 103 Gladstone Street, Birmingham, UK	

The main parts of BBJ's business are manufacturing, distribution and construction. BBJ has been doing business in many towns and cities in Japan for over 50 years.

But today something very shocking has happened.

Scenario Design

This scenario was designed to bring managers into a rather different kind of crisis situation: one which arises from a natural disaster, rather than from insecurity and war. An overarching theme for these Japanese scenarios is, therefore, the extent to which the business skills needed to operate effectively here are similar to those that might prove useful in, for example, a war zone.

As in a conflict situation, there is a serious debate to be had about the business motive for intervention. There may be those who argue that BBJ had no proper role in providing emergency assistance, many who feel there was a moral imperative to do so and perhaps others who focus on the longer-term benefits for business in this sort of initiative.

This idea is at the heart of Pentalemma, that a business is a part of the community in which it operates and that the welfare of that community is of direct importance to that business. This is grounded in the assumption that positive-sum solutions are often of benefit to business and stakeholders alike. In this sense, the central dilemma set up by this scenario—that between profit and philanthropy—can be seen to be misleading. Yes, BBJ was right to intervene in any case; but having done so, are there ways in which it can turn this into a win-win for itself and the community in Kyushu?

The options presented by Pentalemma to the managers provide them with an opportunity to explore the reactions outlined above, which range from immediate withdrawal to full support regardless of cost. In doing so, they will need to consider such questions as reputation management, the return to normal business and the welfare of their own workforce. They should spot the fact that partnership with other agencies—some of them, possibly commercial rivals, can bring quick synergies to a desperate situation.

It is a complex and challenging setting that raises both ethical and business questions. The two are inseparable, and the shrewder management teams will be able to identify a strategy that reconciles them for the benefit of all.

Multi-hazard Disaster

This morning, a huge earthquake and tsunami hit the northern coast of Kyushu, the most southerly of Japan's large islands. It is very serious, similar to the multi-hazard disaster of 2011 known as the Great East Japan Earthquake. Infrastructure has been severely damaged, so that "business as usual" is not an option. Your task is to assess the situation and determine whether BBJ is making sound decisions. You will join BBJ board meetings. Finally, you will report to the CEO of your company, Pentalemma Inc.

You have learned that BBJ acted quickly. It has sent help to the most seriously damaged area within hours of the disaster. There are nuclear power plants there so that radioactivity could be a problem. Multiple aftershocks should be expected and could cause another tsunami. BBJ's manager on-site is Head of Operations Kelvin Reed, a professional who knows how to make decisions. He has acted unilaterally in deploying BBJ's lorry fleet and supplies to run emergency provisions straight into the disaster area. In doing so, he has deployed BBJ staff on a voluntary basis. He has also worsened an already precarious logistical situation for BBJ, in that even without such an initiative, the company would have been struggling to maintain its normal business following the quake.

He has now convened an emergency Board meeting and sought the advice of Pentalemma as to planning options.

Reed has asked you to summarize the situation for him and the team. You have to read many reports to prepare for the meeting. Reed's immediate dilemma concerns the extent to which BBJ should remain engaged in disaster relief. He also needs to secure retrospective approval for the action he has already taken. There seems to be a direct trade-off between profitability and continued philanthropic activity.

Acting Out the Scenario

Description

During this and future meetings, managers will wish to consider whether, in an emergency of this scale, a large company that is in a position to help should simply do so. Beyond that, there are a number of potential future postures ranging from immediate disengagement to continued full support. The model for any such support might also be important in terms of reputation and profitability. What may have begun as a straightforward humanitarian judgment therefore quickly becomes much more complicated.

A set of difficult decisions requires a discussion of several topics:

1. Straight philanthropy: the extent to which a company should act without concern for profit.
2. Crisis partnerships: the value of working with commercial and other organizations to find the best fit in an emergency.
3. Emergency action: when constraints such as the rules of governance may need to be temporarily set aside.

The issues include natural disasters, immediate threats to profitability, workforce safety during an emergency, tensions with the state and other actors, and public relations.

Questions to Consider

1. How should Kelvin Reed's decision to intervene immediately in the Kyushu disaster be justified to the Board?
2. What risks might arise as a result of building partnerships with other commercial organizations during such a crisis?
3. Is pure philanthropy good for business?

There are several significant risk factors to be considered.

Ethical: As a household name in countries around the world and as a respected brand with a long history in Japan, BBJ's assistance to people in the affected areas could make a significant difference. The responsibility to help is, therefore, a moral one.

Technological: The reactors of one of the nuclear power stations of Northern Kyushu have been damaged. In case of a nuclear emergency, BBJ must take all possible measures to ensure the safety of its staff and facilities.

Economic: Due to the crisis, BBJ has been running out of polythene supplies, while raw material prices have shot up, making the company's relief efforts and sales distribution a lot more difficult.

Demographic: Thousands of people are moving away from the zone around the power stations and the affected areas, and an organized evacuation is underway. This is placing additional strain on the road network and disrupting delivery schedules.

Social: Although BBJ has managed to deploy teams of volunteer staff with minimal disruption to normal working patterns, there are health concerns associated with working in such a high-risk environment, which have not been thoroughly addressed.

Some BBJ managers will want to know how much help the company should continue to give. They may feel that the contribution has already been huge. They will recognize the corporate responsibility to provide aid in a crisis situation, but they also know that a business must make a profit. Does giving more help mean earning less profit? Is it that simple?

Some may argue that BBJ should not help, while others feel that helping is the only ethical course. Still others may focus on the long-term business benefits of helping in the short term.

What the managers are really trying to determine is whether it is ethical for companies to do volunteer activities in disaster situations if it reduces, or even eliminates, their near-term profit.

There is a range of options. At one end, BBJ could continue to offer full support. At the other end, they could decide to stop immediately. If BBJ continues to help, what is the best way to do that? The company has to think about its public image and what people say about it. It also has to think about its shareholders and making a profit. Reed's immediate action was a basic human response, but future decisions will be much more difficult.

Among the issues on the agenda will be BBJ's profit, the safety of employees, the company's relationships with local and national governments, public relations and the interests of all stakeholders.

Basic Ideas

The basic idea of corporate social responsibility (CSR) is that a business is part of the community, so it must act as a member of the community. This is sometimes also called corporate citizenship, which means that companies are like citizens of their communities with rights and responsibilities. What is good for the community is also important for the business. By looking at all possibilities, BBJ can make a plan, discuss it with everyone and find solutions that are valuable for its shareholders as well as its other stakeholders. This means that it is possible to make a profit and be a responsible corporate citizen at the same time.

When the Kyushu disaster hit, Reed did not wait. He responded immediately. Will the board support his decision? How should he explain it? Should BBJ work with other business organizations to offer more help during the disaster? What is the risk of doing that? Is pure philanthropy good for business?

BBJ has already decided to send its products, equipment and volunteers to the disaster area. But what happens next? Managers need to think about how the company benefit its many forms of stakeholders as well as how to get back to normal business.

Emergency Field Report

To prepare for BBJ's board meeting, review the below Emergency Field Report sent by Kelvin Reed.

Dear Everyone,

As you know, a huge earthquake and tsunami hit northern Kyushu this morning. It was 9.0 on the Richter scale. This is the largest earthquake ever recorded in Japan. It seems that tens of thousands of people were killed. Even more people are now homeless because their houses were destroyed. The disaster is as bad as the 2011 Honshu earthquake, or maybe even worse. This is a real emergency.

Many people have been affected by the disaster. Since BBJ is a large company in Japan, I believe it is natural for us to help those people. I have offered our help to the Japanese government. However, at this time, the government has just begun planning. That means we should have our own plan. We should do as much as possible by ourselves.

There are three nuclear power stations on the northern coast of Kyushu. NHK reports that at least one of these power stations may be damaged. So, there may be a nuclear emergency in addition to the earthquake and tsunami. BBJ may have equipment that can be used if there is a nuclear accident. I am checking to find out whether we have such equipment.

We can provide supplies such as blankets, food and water. We can also help to deliver those supplies to places where they are needed. Please think about other ways we can support. We must decide on an official BBJ position to explain to the public what our role is. Please reply as quickly as possible so that we can start these activities.

I have set up an emergency meeting of the BBJ board of directors. The meeting will be tomorrow morning at 10:00 Japan time (01:00 GMT). A Pentalemma representative will join the meeting and give us a report on the situation. Someone from the BB Group head office should also be there. If they are unable to join in person, we can set up a conference call.

Earlier today, I spoke with the Minister of the Interior. I believe the Japanese government must include BBJ and other major companies in its plans to help people affected by the disaster. I explained that we, too, really want to play an active role here. I will tell the board what the minister said.

The earthquake and tsunami have damaged BBJ's business activities in Japan, especially our manufacturing and distribution. I will send a separate email within the next hour about these problems.

It is not possible to do normal business immediately after the disaster. We may be able to do so after the situation improves a bit. At the same time, we should consider how much we are actually able to help. We should think about what we can continue doing for the coming months and years, not only for a few days.

I am asking various people for information and advice. They will discuss with us which way to go. After I talk to them, I will share the information with all of you so you can tell me what you think.
Kelvin

Information and Advice

Now, to understand the situation, you must make sure that you have all the information you need. Review the following Country Report and Special Report from Rapid Response Inc. (RRI), one of the world's leading experts on business continuity.[2]

Country Report

RRI has sent Pentalemma a Country Report about Japan. It was written by RRI Senior Consultant, Connie Bernstein.[3] She also sent the Special Report, which Reed asked for as soon as the disaster hit.

Rapid Response Inc.

At RRI, we believe that companies can survive—and grow—in almost any environment if they are prepared. Our Country Reports provide basic information that is easy to understand. We also offer very detailed reports if more information is needed.

Summary of Country Report: JAPAN

By Connie Bernstein

Business Overview

Japan is a leading economy. It is a very good place for doing business. The economic problems of the 1990s have not returned since then. Japan has recovered, and the economy is moving in a positive direction. However, it is important to understand that Japan is unique in some ways. You will see some examples in the following tables.

Japan at a Glance

Economy

GDP	$5,984 trillion
Growth	0.2%
Inflation	0.8%
Unemployment	4.4%
Credit rating	Aa3
Currency	Japanese Yen (JPY)

People

Form of government	Democratic constitutional monarchy
Ethnicity	Japanese 98.5%, Korean 0.5%, Chinese 0.4% and others 0.6%
Religion	• Shintoist 84%, Buddhist 71%, Christian 2% and others 8%. • NB: Many embrace more than one religion or practice none. • No particular tensions; secular state.
Population	127 million

Infrastructure

Internet	Excellent
Roads	Excellent
Rail	Excellent
Air	Excellent
Power	Good
Others (shipping)	One of the largest commercial fleets in the world.

Background

Geography	• Island chain of 6,852 mainly mountainous islands in the Pacific to the east of Korea. • Total area 377,915 sq. km (slightly bigger than Italy). • Japan sits on the junction of 4 tectonic plates; 20% of the world's earthquakes occur there. • Over 3/4 of the population lives in cities on the coasts of Japan's four major islands.
History	• Characterized by periods of expansionism and isolationism. 1605–1854: isolationism following limited contact with Europe. Imposed trading treaty with the USA sparks swift industrialization. • Victory against China and Russia facilitates the Pacific empire. Defeat in World War II; then rapid reindustrialization and growth. • "Lost decade" in the 1990s. • Massive earthquake and nuclear disaster on Honshu in 2011.
Natural resources	Negligible

Japan has a democratic form of government. The court system is independent. There is freedom of speech and freedom of the press. The government is generally effective and efficient, and its officials are usually honest.

Economy and Business

The modern Japanese system of business and the economy functions quite well. This could mainly be because there is very little trouble between companies and labor unions. It is easier for businesses and workers to agree than it is in many other economies. The two sides communicate well with each other. It is sometimes difficult for people from other countries to understand how this is possible.

Japan is still an economic superpower like the United States of America and China. It is excellent in the high-tech industries, especially electronics, robotics and telecommunications. There is a high level of investment and research. This is one of Japan's major strengths. Japanese companies are usually very good at long-term planning.

The Japanese economy was weak in the 1990s, partly because prices for property and shares dropped to very low levels. This was a shock for the nation's leaders as well as for the Japanese people. The economy started growing again because of improvements to the banking system and the rise of China, but only very slowly.

More recently, the Japanese government's policy of "easy loans" seems to be working, and the economy continues to grow. However, Japan's national debt has long been over 200% of its GDP. The population is aging fast. Environmental issues and people's concerns about these issues make it difficult for some companies to continue doing business.

Natural Resources

Japan imports over 90% of its resources, including 60% of its food, all of its coal and nearly all of its oil.

This is a big weakness, but Japan has even bigger strengths. Those strengths are its industries and service businesses, which are always developing new and improved products and services.

Security

Modern Japan is a safe place in which to live, work and do business. The crime rate is low (although organized crime has been a problem, as it is in many other countries). There were violent protests in the 1960s against the building of Narita Airport. But there have been very few such major protests since then.

Japan's international relations could deteriorate over the coming years. China is growing more powerful, and there are some arguments with China concerning several small islands. Nobody knows what North Korea will do, and South Korea is becoming prickly.

Current Risk for Japan: Low

Generally, Japan's business future looks bright. The business culture is sometimes difficult to understand, but there are many excellent opportunities. Clearly, one must never forget that earthquakes will continue to occur.

RRI Special Report

Business and Crisis: The Kyushu Earthquake

Introduction

The northern coast of Kyushu has been struck by a major earthquake. The quake caused 30-meter-high tsunamis to hit several nuclear power plants in the area. The disaster is terrible and very complicated, similar to the 2011 disaster. RRI's experts are assessing the risks and options for business in Japan.

The disaster in Kyushu has caused problems in many parts of Japan. These problems are a direct threat to many businesses. Businesses will want to do something to help those affected by the disaster. But there is a risk that some businesses will do more than their investors would like. There may be some chances for increased sales in some industries for reconstruction in the short term, whereas in the long term, there will be opportunities and challenges.

The Current Situation

The following is a brief report on the situation several hours after the earthquake hit. Many thousands of people are dead. Even more people are homeless. Only emergency vehicles are allowed to use the roads and railway lines throughout Kyushu and much of Honshu. Businesses cannot use them yet. They could send their goods by sea, but that would be very expensive. Distribution across Japan is, therefore, a very serious problem.

Some businesses have acted very fast to provide help to people in trouble. There have been many TV and newspaper stories about multinational corporations (MNCs) sending trucks loaded with food and blankets. In some areas, the Japanese government has borrowed cars or trucks and received some goods directly from businesses.

In Kyushu, many factories outside the area that is directly affected are closed. Due to the risk of radioactivity, the government is blocking access to the area around the three nuclear power plants.

To summarize, the crisis situation right now has created major difficulties for many businesses in Japan.

Short-term Threats

Many companies will have problems with distribution, sales, staffing and safety. Some will have more serious problems than others. It depends on where they are and what type of business they are doing.

Naturally, businesses would like to help people in the affected areas as much as possible. However, they must be careful. They have to think about the impact on their companies. Reasonable and careful CSR activities, on the other hand, can be very good for a company's image. It is often difficult to find a good balance.

Short-term Opportunities

Some businesses may benefit from the earthquake and tsunami, as many people saw in 2011. For example, there will be a high demand for businesses that specialize in various aspects of the cleanup process. There will also be a strong demand for alternative transportation and large-scale equipment. We recommend a sensitive but practical approach. If a company is able to offer good quality at a fair price, it should not hesitate to do so. As mentioned, controlled CSR efforts will also benefit corporate image.

Long-term Threats

It is difficult to plan the next steps when new disasters can occur at any time. In addition to the suffering and death they cause, such disasters also destroy assets worth billions of dollars. During this very sensitive stage, there is a serious risk that the public may lose respect for a business that they feel is doing too much or too little. Therefore, businesses should watch the economic situation carefully.

In particular, nuclear power plants may have to be shut down. This would mean that Japan must import almost all its energy. That would become a huge cost for the economy.

Long-term Opportunities

Obviously, a major reconstruction program will be needed for Kyushu. Industry should begin planning now for the best way to take advantage of business opportunities created by the reconstruction program. Towns and cities have been badly damaged or destroyed. That means policymakers have to find ways to increase business activity and jobs in those areas. They will probably offer money and other incentives to people to live and work there. They will offer something to businesses as well. If a company wants to establish a new factory or service center there, that may be a good business decision.

Conclusion

As with all fast-changing situations, the tragedy in Kyushu presents threats and opportunities. In 2011, some companies went bankrupt due to the events. Other companies benefited. Often, the difference was based on their ability to seize opportunities and manage risk. This will be important this time too.

Contact RRI if You Need More Details

Rapid Response International employs consultants who are experts in all regions of the world. For research into specific risks, situations or crisis areas, the best decision a modern business can make is to ask for a special RRI Report. Our reports are always recent and written just for your business.

Devising the Pentalemma

Having read these two RRI reports, you can begin thinking about the range of options open for consideration by BBJ managers.

This is truly an emergency situation. The island of Kyushu is at the center of a huge disaster. Many lives have been lost. Many buildings and infrastructures have been damaged. Because of the tsunami, flooding is a threat to the nuclear power industry along the northern coast. BBJ immediately sent its trucks to the disaster area, carrying food and various other goods. This was some of the first help to arrive. Clearly, profits will be affected.

One obvious question is whether or not BBJ should continue its disaster relief activities. According to company rules, perhaps Reed should have asked the board before using BBJ's resources. Using those resources could have a negative impact on day-to-day business, so it may be more responsible for the company to stop using them.

With the aim of identifying at least five potential solutions, you are calling cessation of relief activities "option A."

If BBJ continues to offer disaster relief, what kind of support should it be, and what levels are realistic? Continued full support is one option. That would be "pure CSR," or philanthropy, which you call "option B."

Your research so far shows that some BBJ managers have taken notice of a recent research report arguing that the public does not think CSR is always necessary for companies to be successful. However, the reconstruction program for Kyushu may create big opportunities for BBJ. That means a certain level of relief work may be a good investment.

As a Pentalemma manager, you should be prepared for a deep discussion with the Board. How will Japanese society likely react to BBJ's decision? In particular, how will the people in northern Kyushu react? Will future sales be affected? Will BBJ's profits be affected?

Any decision brings with it some risk of negative TV or newspaper coverage, which could be very bad for the company's image, not only in Japan, but also worldwide.

Your challenge is to create a strategy that satisfies BBJ's interests and helps the communities that have been affected at the same time. You also have to explain this strategy very clearly to the other people at the meeting. To do this, you will need to consider the advantages and disadvantages of options A and B as well as the three alternatives detailed below: options, C, D and E.

Option C—Respect the Rules. Slow Down so that Everyone Can Get on the Same Page

There are special cases in which it is more important to take action than to follow all the rules. For example, in contrast to normal situations, there may be no time to ask for permission to do something.

Reed felt it was best to make an independent decision immediately after the earthquake. Some of his actions did not adhere to BBJ's rules. It may be argued that rules cannot always cover every possible event. When there is an emergency, managers must be free to do what they think is best. Following the rules is very important, but it should not prevent managers from making critical decisions during a crisis. After that, managers must be able to explain their actions.

Now may be the time to do just that. In addition, it may make more sense to deploy resources toward acquiring a deeper understanding of what constitutes effective and sustainable assistance.

Option D—Smarter Relief

BBJ could continue to offer philanthropic support if the company's decision-makers think that is the ethical action to take. Also, if BBJ goes the other way and reduces its relief activities, its image may be damaged. The damage to the company's image could affect sales and even the share price.

Philanthropy may be good for BBJ's image, but sales may drop due to transportation problems, and costs may rise. Therefore, relief efforts must be sustainable, and the company should find partners so they can do these activities together. Disasters create many changes, and changes create opportunities. BBJ should look for new relationships and find creative solutions to the problems they face. In this way, both the community and the business can benefit longer-term.

Option E—Partner with Other Stakeholders

When there is an emergency or crisis, working with others who are involved can help to have more impact for less cost. It also creates possibilities for future collaboration. By sharing the burden of relief and reconstruction work, BBJ protects its business and its image. The people in the area, as well as the businesses, will appreciate BBJ as a friend, not only in good times, but also in times of crisis.

Conclusion

The Pentalemma approach requires solutions that benefit all stakeholders. To do this, you have to understand the different stakeholders' opinions and help them develop sustainable agreements. BBJ has some stakeholders who like the idea of pure philanthropy. They feel it is more important in this situation, and they believe that continued philanthropy is an ethical business decision.

But there are others who think BBJ's profit is the highest priority. They see philanthropy as an unnecessary expense. Your challenge is to support BBJ so that the management team can make the best decision for themselves and their stakeholders. Reed trusts Pentalemma, and he is counting on you to give him and his colleagues good advice. With five options on the table, you have given the client your best advice. The management team now considers the five options and your considerations regarding each one. This includes the suggestion that the optimum outcome will be a win-win situation for BBJ and all of its stakeholders.

At this point, let us fast-forward to consider five hypothetical sets of consequences.

Initial Impact on Reputation

Option A—Cessation of Relief Activities: Tokyo Residents Organize a Boycott of Brighter Brand Goods

Sachiko Yosano is a Tokyo housewife who was a regular BBJ customer. But when she heard that BBJ had ceased its food deliveries to the stricken earthquake zone, she felt she had to do something. "They say they need their lorries to deliver their products to Tokyo. Well that's ok; they won't need so many now—we will stop buying their goods!" Mrs. Yosano has since orchestrated an online campaign which has garnered 158,000 "likes." BBJ was unavailable for comment.

Option B—Pure CSR: CEO Rolls up His Sleeves

Our picture shows Kelvin Reed, Chief of Operations for BBJ, helping to unload a lorry in the earthquake zone. Mr. Reed was too busy to comment when quizzed by our reporter. However, residents told us that lorries from BBJ were among the first on the scene; with blankets, bottled water and canned food. A company spokesman later said that Mr. Reed had wanted to spend some of his weekends at the site but that he was needed back at their Nagasaki headquarters to co-ordinate the effort. We hope he doesn't get into any trouble!

Option C—Respect the Rules: Have They Had Enough Already?

On tonight's Channel 5 special report, Yoshi Yoshiko follows business leaders as they work with Non-Governmental Organizations (NGOs) and the government in Kyushu's stricken earthquake zone. But most of the day was spent in a UN-convened "Industry Liaison Group" meeting. Are these people fiddling while Rome burns? Tune in at 7.

Option D—Smarter Relief: Industry Settles In for the Long Haul

It is refreshing to see that industry appears ready to continue with its backing of the Kyushu earthquake relief effort. Although the sight of

BBJ and KZ Food lorries rumbling toward the area has become less common on our screens, these giant companies are still very much engaged. As Mr. Reed explained to me, the relief effort has simply become a lot "smarter." "You'll find KZ lorries carrying BBJ products in some parts of the quake zone. I never thought I would see that, to be frank. But it just depends on what seems the best fit at the time." For the people of northern Kyushu, these are probably academic distinctions; it's good to know, though, that these companies are still doing it because they think it's right.

Option E—Partner with Other Stakeholders: Telecom Japan and BBJ in Quake Alliance

Strange bedfellows they may be, but Telecom Japan and BBJ have struck up a remarkably co-operative relationship in the days since the earthquake hit Northern Kyushu. "I just picked up the phone to Yoshi (Yoshi Nimura, CEO, Telecom Japan) when I realized we needed some technical help with our Website," said Mr. Reed. "We were getting a thousand hits a minute, and the lines suddenly went down. What I didn't realize was that TJ (Telecom Japan) were having problems getting their people into the quake zone. Using our fleet, we just took it from there." According to diplomatic sources, this kind of partnership, which also includes NGOs and the government, has huge potential. "This is something almost revolutionary in terms of crisis management," said US Ambassador Nick Earnshaw. "I'm hoping to ensure we carry it through and learn for the future."

Later Impact on Reputation

Option A—Cessation of Relief Activities: Media Reaction

Brighter Brand Japan (BBJ) has been roundly castigated for the cessation of its emergency supply runs early yesterday. The company finds itself the center of a media storm—forced onto the defensive and losing most of the goodwill generated by its early response to the disaster.

Option B—Pure CSR: Media Reaction

Footage of BBJ's CEO unloading food aid amidst the ruins was worth millions in terms of PR. His refusal to comment on the effort underlined the altruistic intent of the company and also played well. Being juxtaposed with dithering politicians reflected poorly on national leaders, but can hardly be blamed on industry. This has all been a good news story from a selfish BBJ perspective, whatever the motives.

Option C—Respect the Rules: Media Reaction

The establishment of the Industry Liaison Group—effectively a chance for the multinational corporations to reflect on their next moves in this crisis, has not really registered with the media. Instead, the focus has been on the nuclear emergency, and the shortcomings of that industry and the government. Things may change if the press scents some kind of charity fatigue amongst the big donors.

Option D—Smarter Relief: Media Reaction

"Japanese Defense Force delivers BBJ baby food" made the front page both in Japan and in BBJ's home base in Britain. It was a lightweight news story after so much bad news, but positive nonetheless. The more sophisticated and efficient systems that have been established in recent days, many of them, based on such partnerships, have been mostly hidden from view.

Option E—Partner with Other Stakeholders: Media Reaction

It has been difficult to sell our partnership-based aid model to the media to date because it is too early to see very clearly the benefits the aid model will generate. "Business Today" carried an interview with Mr. Reed and Yoshi Nimura (Telecom Japan), but it came across as rather theoretical.

General Impact on Firm Value

Option A—Cessation of Relief Activities: Scaling Back

COMPANY REPUTATION: When BBJ left the areas of operation depleted by the earthquake and subsequent tidal waves, their position was viewed by many of the locals as cutting and running. BBJ's reputation, therefore, suffered from this choice as companies who stayed put and helped with the relief efforts have achieved a collaborative working relationship with the local government and the NGOs on the ground. By shutting down operations in the affected area, sales were down, and shareholder reaction was negative. The share price is down almost 25% over the last year. The overall impact is estimated to be $5,000,000.

SALES: Domestic sales are down overall for the year. This is in part because BBJ decided to shut down the three factories in the devastated areas and also because they were viewed as deserters when the regions that they operated in needed them most. While we hoped to take advantage of the commercial opportunities that might arise after the rebuilding process begin, it was not enough to combat our decision to leave the areas unattended. The net value of sales is now only worth $23,000,000.

RISK MANAGEMENT: BBJ has reconfigured its entire risk management strategy in light of the crisis in Japan. Obviously, investment has been required in areas such as physical security, but we have also had to devise plans, which might avoid such a sales collapse during any future episode. Our "wargaming" now envisages a much bolder involvement with relief efforts, as it seems that the posture taken to date has actually increased the risks facing the company. The online product boycott launched in Tokyo was devastating. We infer that the net value of these new risks costs the company $5,000,000.

SAVINGS IN SUPPLY COSTS: There were no savings in supply costs. The earthquake destroyed major roads and supply routes. The raw materials we use for our products have become scarce due to the rebuilding efforts, and their costs have skyrocketed over the past year as well. The estimated savings for the year were wiped out by the earthquake and the damage it did. We suspect that BBJ took a loss in supply costs this year to the amount of $7,500,000.

STAFF MOTIVATION: Staff remained motivated throughout the initial relief efforts but became wary after seeing what BBJ's reaction was to the crisis. Some also feared the effects of nuclear fallout and requested to be sent back home. This crisis caused staff motivation to dwindle in the three of the seven total sites. The net value effect of this was reduced to $5,000,000.

Option B—Pure CSR: Full Philanthropic Engagement

COMPANY REPUTATION: By taking the corporately responsible route and staying to help with the relief efforts, Brighter Brand Japan now has favorable relations with the local government and NGOs. The press coverage has been superb. Although normal operations have not yet resumed, the share price only took a small hit due to decreased sales and increased costs. However, that was the least of our concerns as we rushed to provide support to the affected area of northern Kyushu. The net value of the company's reputation is $20,000,000.

SALES: Our logistics capabilities were down at the three plants that operate in the region affected by the earthquake. Therefore, sales were also down as production fell, and supplies and employees became scarce during the ensuing relief efforts. Thankfully, we are now approaching full operational status despite continuing relief work. The net value of sales this year was $29,000,000.

RISK MANAGEMENT: BBJ has had to recast its risk profile and appetite in recent weeks. Our experience has shown us that for the future, we need to plan on significant staff abstractions and loss of "business as usual" if BBJ is to respond in this manner to national crises. The corollary of that analysis is multi-faceted: from holding larger stocks of our products to ensuring proper psychological support to staff. All such precautions come with a price tag. This kind of full-on CSR posture brings a great deal of risk management with it. BBJ's prior years' estimates were not met, and more precautions led to higher expense for managing our risk. We estimate at least a $20,000,000 net loss to date.

SAVINGS IN SUPPLY COSTS: Estimates of the loss of savings in supply costs are high. Major infrastructure damage as well as inoperability both contributed to BBJ's decreased estimates for savings in supply costs. Many of the raw products we use to make our product were redirected to the relief effort, and rightly so. With months of downtime, we were unable to benefit from economies of scale and our purchasing power diminished as the price of our raw materials grew exponentially. All told, the net value of our losses was $15,000,000.

STAFF MOTIVATION: Burnout and safety were the biggest concerns BBJ had with regard to employees. When the disaster first struck, many of them were volunteering around the clock, and we had more than 91% of our staff present at any given moment. We brought in agency workers for support, but that did not fully resolve these issues. As time passed and normal operations resumed, morale and motivation improved. It seems, however, that BBJ may have been asking too much of its people and that a slightly more balanced response might have been wiser. The net value of staff motivation is currently $10,000,000.

Option C—Respect the Rules: Pause and Plan

COMPANY REPUTATION: Partnering with local agencies, NGOs and the Japanese government to hold weekly meetings to determine how best to proceed were viewed with a strong measure of cynicism by the media. They viewed our decision as a stall tactic, and people believed them. Although BBJ was acting in the best interests of the people of northern Kyushu, our actions had a negative impact on our reputation overall. The net value of this reputational hit is thought to be a $5,000,000 loss.

SALES: Although operations returned to normal in a timely manner, sales were still down for the period as production dropped for some time at the sites in the earthquake-affected region. Any increase in sales for the perceived goodwill of our efforts was to a degree offset by this. Sales net value is estimated to be $43,000,000.

RISK MANAGEMENT: The pause in support activity brought about by the launch of the Industry Liaison Group provided BBJ with an opportunity to consolidate. Some risks were thereby mitigated—especially in terms of staff absence and fatigue. We estimate significant savings in terms of avoiding future health issues, such as post-traumatic stress syndrome—which BBJ could be accused of having engendered. Our net value for risk management is consequently valued at $8,000,000.

SAVINGS IN SUPPLY COSTS: The estimated savings in supply costs were not met during this period due to the disaster. First, BBJ used some of its supply infrastructure to bring relief to towns and villages without food or water. Second, major infrastructure damage across the region meant we had to fend for ourselves in order to ship our product. Third, commodities that we use in our production have been in short supply, as the authorities have priority where they are needed for relief efforts. In summary, nearly $15,000,000 was lost.

STAFF MOTIVATION: Brighter Brand Japan's staff have been extraordinary during the relief efforts, with nine out of every ten employees showing up every day to aid and assist the NGOs and other organizations involved in the cleanup and repair of northern Kyushu. Although some employees suffered burnout, because we effectively paused our support in order to reconsider commitment levels, most were quickly able to return to normal working patterns. Staff motivation had a net value of $13,000,000.

Option D—Smarter Relief: A Mixed Economy Approach

COMPANY REPUTATION: The more sophisticated relief model we have put in place has attracted little public comment. Our shareholders also seem content. This decision has allowed Brighter Brand Japan to get on its feet quicker and keep sales figures relatively even. We estimate our company's reputation value added to be worth $17,500,000.

SALES: Sales barely dipped during the current reporting period. We were able to transfer many of our initial responsibilities back into the hands of local government and NGOs after only a few weeks of assistance. Although our relief work has continued, the way in which we deploy assets is much more sophisticated and therefore more efficient. At the same time, our buoyant reputation and improving the ability to maintain production have supported sales. We estimate that using this strategy rewarded the company's sales with a net value of $48,000,000.

RISK MANAGEMENT: The significant risks that faced the company were mitigated early on. Perhaps the most difficult of all the risks was how we would get our products into the hands of customers. How could we deliver supplies to the doors of those in need? The company was also concerned about the negative media reaction and consequent PR related to turning relief operations back over to the NGOs and government. Risk management had a positive net value of $10,000,000.

SAVINGS IN SUPPLY COSTS: By burden-sharing in the critical area of transportation, we were able to secure what surely would have been millions of dollars in lost savings had we not pursued this option. Although BBJ still struggled with its own distribution problems, the net value of savings was estimated at $12,500,000.

STAFF MOTIVATION: Employees were highly motivated from the outset to help with relief efforts. Fortunately, we were able to put many of them back to work at their normal jobs because production came back online so quickly. They were also encouraged by seeing that their company stayed put and helped in reconstruction and disaster relief. Essentially, this balanced strategy has allowed BBJ to deploy staff to the best effect, thereby ensuring their continued support. The motivation of our employees had a net value of $10,000,000.

Option E—Partner with Other Stakeholders: Partnership-based Commitment

COMPANY REPUTATION: The Company's decision to partner with other multinational corporations seems to have benefited BBJ's reputation. Although press coverage of this aspect of our work has been subdued, our monitoring indicates a sense of appreciation among the public. By working together with everyone from our competitors to the UN and NGOs, our first-priority commitment to disaster victims has become tangible. Our rapid intervention, along with these other entities, showed that Brighter Brand Japan cares about more than just profits and losses and has a genuine concern for the people and their families in the regions where the company operates. This decision added a net reputational value of $30,000,000.

SALES: BBJ's involvement in the reconstruction efforts and, most importantly, getting the supply chain moving again, has meant that sales for BBJ's products are back in line with estimates. Partnering has enabled us to reestablish our operations quickly, and our popularity may have boosted sales already. The net value of these sales benefits was $44,000,000.

RISK MANAGEMENT: We've found it possible to bring our partnerships to bear on risk management as well as immediate crisis relief. Already, we have outlined ideas for risk-sharing with some of the corporates we have been working with in Kyushu. For example, BBJ has been in a position to provide Business Continuity transport assets to Telecom Japan, in return for IT resilience support. If now, why not during any future emergencies? Already, BBJ has earmarked notional savings here. The risk management net value is estimated at $7,500,000.

SAVINGS IN SUPPLY COSTS: Although BBJ suffered like everyone else from the effects of the northern Kyushu earthquake, logistics were up and running within a week. This left plenty of leeway for our prior year estimates of expected savings in our supply costs. Partnering with other companies and the government, we were able to receive raw materials for BBJ projects as well as get the completed inventory out the door and to its final destination. These savings amounted to a net value of $10,000,000.

STAFF MOTIVATION: People want to work for an organization that they know is caring. Suppressing ego and working together with other companies, even competitors, showed the workforce that BBJ is on their side. In fact, we have been flooded with applications from people all over the world who want to work with a company that promotes corporate social responsibility. Sickness and absence have tumbled; productivity is up 13%. These benefits have a net value worth $12,500,000.

Reflecting on Chapter One

The Scenario in Context

Pentalemma consultants in this scenario were tasked with advising Brighter Brand Japan (BBJ), part of a large multinational corporation with a well-established manufacturing, distribution and civil engineering business across Japan. This was truly an emergency situation, engaging Pentalemma in the first hours after Japan had been struck by another extremely powerful earthquake. In an echo of the Great East Japan Earthquake of 2011, the island of Kyushu was now at the center of a huge disaster. Widespread flooding had threatened the nuclear power industry along Kyushu's north coast.

BBJ had immediately deployed its vehicle fleet, carrying stocks of water, food and manufactured goods in support of the relief effort, representing some of the earliest help to arrive. Clearly, profitability would be affected, and the company now needed to consider its options.

Dilemmas and Challenges

The likely starting point for management teams in this situation would be to consider whether or not BBJ should remain engaged in disaster relief at all. It seems that the Head of Operations may have technically exceeded his authority in deploying BBJ's resources without reference to the Board. Undoubtedly, such a move would quickly impact on BBJ's day-to-day business. There is, therefore, an argument to be made for immediate withdrawal from relief operations—perhaps based on a reading of the country profile provided by Connie Bernstein of Rapid Response Inc. That document, however, may also contain suggestions as to why pulling out might not be optimal for BBJ.

If BBJ were to continue offering support, the more difficult practical question concerns the form that such assistance might take. Managers had an array of strategies to consider, including one advocating full support, whatever the cost to the business. In a sense, this might be regarded as "pure" CSR, or philanthropy. Managers may well have been asking themselves what business BBJ had with such a stance.

Such discussions would have led to other aspects of the situation in which BBJ found itself. There were important considerations beyond those required for daily operations. What are the characteristics of Japanese society, and how might these influence reactions to the various options on the table? As a subset of this, there may be different views held within the community most directly affected, i.e., the residents of northern Kyushu. Might this affect future sales or growth? As a corollary to this, 24/7 global media attention should also feature in the management team's thinking. Getting on the wrong side of the media argument in a crisis of this magnitude can have a serious impact on company reputation, locally and globally. There may even be opportunities here for future development, which could be costed in, albeit in broad or speculative terms.

This scenario was different from those usually presented to Pentalemma consultants, in that their problem was the crisis itself, rather than simply being a dilemma, or dilemmas, arising in a chronically difficult environment. Teasing out a strategy that protected BBJ's interests yet attempted to benefit a stricken community was, therefore, no easy matter.

Notes

1. The roots of the pentalemma perspective can be traced back as far as Aristotle and his square of opposition (Béziau & Payette, 2012). While it only defines four

points of opposing propositions, a fifth mid-point of compromise between these can be inferred. The German philosopher, Johann Fichte, introduced the thesis-antithesis-synthesis triad in his 1794/1795 publication, Foundations of the Science of Knowledge (Breazeale, 2013). This triad implies a process of reasoning that develops and then reconciles opposing views. Another German philosopher, Wilhelm Krug, made explicit mention of a pentalemma in 1827 in his dictionary of philosophical terms, defining it as a five-dimensional problem or "five-horned conclusion" (*ein fünf-gehörnter Schluß*).Between the 1950s to the 1970s, Norwegian sociologist and pioneer of peace and conflict studies, Johan Galtung, developed in his lectures and unpublished books a five-point model for analyzing social conflict. These were later published in the Encyclopedia of Violence, Peace and Conflict in 2008 (Kurtz, 2008). I had the opportunity to join many lectures and workshops over a period of several years in the 1990s in which Galtung would demonstrate how to analyze conflicts with the model. At a lecture he once gave in Germany in 1999, he drew the model on a flipchart while telling the students, "I don't understand why more people don't use this." I took note and began experimenting with it from that day. Galtung was the first person to publish analyses of social conflict from a five-point perspective, something he confirmed in a discussion I had with him about this at a conference in 2016 in Japan.In the management literature, Robert Blake and Jane Mouton (1964) developed a five-point model in their book *The Managerial Grid: The Key to Leadership Excellence,* differentiating between leadership styles in relation to a concern for people versus a concern for production. In 1981, Fisher, Ury and Patton then introduced a similar model for interest-based negotiation in their bestseller Getting to Yes.More recently, the model has also been used in the psychology and related coaching literature, as in Thomas Augspurger's 2008 book on constructive leadership. Different versions of the diagram are still used in game theory to depict Nash equilibrium or Pareto optimality. Drawing on these influences, especially Galtung's work on conflict theory, I apply the pentalemma model to the challenge of satisfying the competing needs of corporate stakeholders.

2. Rapid Response Inc. (RRI) is one of the world's leading business continuity consultancies. The company provides advice on maintaining business in volatile and complex environments. Specialized in crisis management, RRI helps businesses secure their operations under extreme conditions such as natural or manmade disasters. Emphasizing foresight and preparation, the company believes that a well-prepared organization should be able to carry out its business in almost any environment. The client base is diverse, including groups like Pentalemma that advise on broader areas of risk, multinationals, governments and individuals. RRI consultants have been recognized experts in their field since 1985. In the RRI reports, a "high" status does not mean simply "stop": it means stop, review and where appropriate, proceed with caution. Between the extremes of "low" risk and "severe" environments, "guarded", "elevated" and "high" assessments may be

regarded as transitional. These show when a country is moving significantly in one direction or another, providing an overview of the security climate at a glance.

3. Connie Bernstein is a senior consultant at Rapid Response Inc., which she joined in 2011. Connie manages a small team of consultants who undertake detailed research on global "hotspots," and works with clients who need bespoke and intensive business continuity advice. She used every opportunity to keep her skills fresh by maintaining her own client portfolio and working on the frontline whenever she can.Much of her time at the moment, however, is devoted to building up the RRI Country Profile Service. This is a new idea, based on the concept of providing up-to-date summaries of countries around the world and what it means to do business in them. This is a growing archive and a product for which the company regularly receives excellent feedback.

Further Reading

Blake, R. R., & Mouton, J. S. (1964). *The managerial grid*. Houston, TX: Gulf Pubs.

Chomsky, N., & McChesney, R. W. (2006). *Profit over people: neoliberalism and global order*. New York: Seven Stories Press.

Fisher, R., & Ury, W. (1981). *Getting to yes*. Boston: Mifflin.

Freeman, R. E., Harrison, J. S., & Wicks, A. C. (2007). *Managing for stakeholders: survival, reputation, and success*. New Haven, CT: Yale Univ. Press.

Galtung, J. (2012). *Peace by peaceful means peace and conflict, development and civilization*. Oslo: Prio.

Gray, J. (1993). *Beyond the new right: markets, government and the common environment*. London: Routledge.

Gray, J. (2004). *Heresies: against progress and other illusions*. London: Granta.

Gray, J. (2009). *False dawn: the delusions of global capitalism*. London: Granta.

Gray, J. (2014). *The silence of animals: on progress and other modern myths*. London: Penguin Books.

Horton, J., & Newey, G. (2007). *The political theory of John Gray*. London: Routledge.

Chapter Two

Strategy and Tactics

As part of a Pentalemma Inc. crisis advisory team led by Sylvia Phillips, you are working on your second assignment to Brighter Brand Japan (BBJ), which is still struggling with various stakeholder issues ten days after the earthquake and tsunami. You and your colleagues continue to help BBJ find creative solutions that satisfy as many of their stakeholders as possible.

As mainstream agencies such as the government and the UN have now grasped the situation, the need for emergency relief from the private sector has diminished. The question facing Kelvin Reed and his Board is about broader long-term engagement. He sees a case for continued involvement, as there is plenty more work to do. An immediate example presents itself in the form of a request for help in refurbishing a tsunami-damaged water purification plant at Saiki, in Oita prefecture. This can be viewed either as a straightforward commercial matter or an opportunity for BBJ to continue its philanthropic work. Whatever BBJ does, it must follow the standards of its parent company, Brighter Brand Group.

The question concerning the water plant is tricky because this kind of activity is normally mainstream business for BBJ. The company has a well-respected engineering arm that undertakes such work across Japan. The absence of clean water is an obvious environmental and health hazard affecting hundreds of thousands of people. The moral and business dilemma is complicated by the role of Civil Solutions (CS). CS is another company operating in this field, and they have offered to do the work on the water plant for free. You have learned, though, that the local authorities have high regard for BBJ's experience in this area and that price may not be the only criterion. The commercial threat that CS poses to BBJ's dominance of this sector in Japan cannot be ignored either.

Any decision concerning the water plant will need to consider BBJ's strategic positioning vis-à-vis the disaster and its overall market in Japan. Indeed, Reed is quick to spot this and introduces his Field Report in these terms. Normal business is not going well. Profits are markedly down in the north of the country. The lorry fleet currently deployed in relief work is desperately needed to restore that situation. Other companies have already pulled out. Should BBJ also stop? Or is it time to change BBJ's support with a new focus on rebuilding? The synergies secured through partnership work during the early days of the crisis suggest a way forward here. Even more broadly, BBJ has an existing growth strategy for Japan, and it needs to consider how this might be impacted by its current dilemmas.

A set of difficult decisions requires a discussion of several topics:

1. Commercial competition and its extension into the arena of CSR
2. The benefits of integrated partnership planning
3. Mixed models for capital projects—public/private initiatives

The many issues raised in the scenario include logistics; water infrastructure; civic leadership; emergency partnerships; publicizing philanthropic work; employers' obligations and disaster fatigue.

Scenario Design

The overarching design challenge with this scenario was to confront managers with a choice about "staying the course." While the first BBJ scenario introduced the notion of straight philanthropy in a crisis, that can be quite an easy call for business leaders—especially in a disaster of this magnitude. It gets harder as the bottom line begins to move into the red, especially when a plausible case can be put that the emergency has passed. In order to point this up, it was decided to devise an immediate choice which would force the company to reconsider its position.

BBJ is at a turning point in Kyushu because it desperately needs those lorries elsewhere and yet it is now asked to do something rather different, namely to become involved in a medium-scale capital project. Some change in strategy must be determined, and this requires consideration of the broader posture. The other ambition here was to introduce the slightly edgy question of competition within a relief operation. Some BBJ managers may not feel entirely comfortable with looking at CS in those terms. This is but another facet of the complicated situation they are asked to resolve. You don't stop being a competitive businessman because you are working for the public good.

The converse of that is the degree to which businesses can set rivalry aside and build partnerships in a crisis. Both of these phenomena have been built into this scenario. Managers should appreciate that as well as keeping a wary business eye open for the competitive implications of their decisions, partnerships can make business sense too. The synergies described in connection with BBJ's partnerships with Telecom Japan and KZ Foods very clearly achieved a greater impact on the ground at a reduced cost for those businesses.

This second scenario in Japan, therefore, poses difficult questions in a highly complex setting. A creative and positive-thinking management team will deduce that continued engagement in what remains a large-scale disaster need not mean commercial suicide. Even so, realistic judgments are required to protect commercial interests and make the most of partnership opportunities. These two points are not necessarily contradictory.

Questions to Consider

1. Were the "disaster partnerships" described in Kelvin Reed's report plausible, or does the real world not work like that?
2. Is it useful, moral or shrewd to consider the behavior of rival firms in a disaster situation in terms of "competition"?
3. What kind of "public/private" funding options might exist for a project such as the cleanup of the water plant?

Significant risks in Chapter Two include:

Economic: As BBJ's staff, donations, goods and lorries are part of the disaster relief effort, it has become more difficult for the company to ensure its sales supply, which is having a significant impact on revenue.

Ethical: The longer BBJ hesitates to decide whether to get involved in repairing the Saiki water processing plant, the longer many will go without water. Over 150,000 inhabitants and 100,000 evacuees continue to be without mains water.

Legal: Although BBJ has been praised for supporting employees who want to assist with the relief efforts, the company needs to be clear about whether this is considered part of their formal obligations to the company.

Social: The relief efforts in which BBJ is involved have resulted in positive media attention; the company needs to be careful not to be misunderstood as exploiting the crisis to boost their public image.

Technological: BBJ has the experience required to repair the damaged water plant in Saiki, and there are expectations for the company's involvement. BBJ, however, is not certain that all of the necessary technology will be available.

The Task at Hand

Your task, as always, is to assess the situation and help BBJ managers make good decisions. You will gather information from Pentalemma and from BBJ. You will contact experts to get their advice, and you will do your own research. You will also communicate with Reed and the other Board members. Finally, you will report to the CEO of Pentalemma. Reed has cc'd you in an email to your team leader Sylvia Phillips.

Japan Disaster: Update

To: Sylvia Phillips, Head of Pentalemma Advisory Team
From: Kelvin Reed, Head of Operations, Japan
Subject: Crisis over—what now?

It has been ten days since the Kyushu earthquake and tsunami. BBJ has been praised for its early response to the crisis. Now we are past the first stage of the crisis. It is time to shift our focus. There is less need for emergency evacuations and food deliveries. The question now is: How much should we do, and for how long? We are a profit-based corporation, and we should get production and sales across Japan back to normal levels as soon as possible.

For example, nearly 25 of our 200 trucks are still being used to deliver disaster relief. Because of that, we cannot provide a regular supply to many of our larger customers. In parts of Honshu, our products are no longer available on supermarket shelves.

Also, many members of our staff are still doing volunteer work in the disaster zone. People appreciate BBJ for our relief activities, but we also need our staff to complete their normal work in the office.

A more specific dilemma is the damaged water purification plant at Saiki. This was damaged by seawater, and 150,000 inhabitants are still without water. The mayor of Saiki called me today, asking whether we could help to repair the plant. I said we would consider it. Normally, the cleaning and maintenance of a water plant would be a commercial activity for us. We have to decide whether to charge a commercial rate, offer a discounted rate or do the work for free.

We really need to decide how much we will be involved in rebuilding the disaster area. Then we can decide how much support we can offer. Meanwhile, we should let the mayor know what we intend to do about the water plant as soon as possible.

I welcome your advice.

Kelvin

Phillips has then shared the following with you as well as the BBJ management team.

Pentalemma Inc.: Briefing

To: Brighter Brand Japan (BBJ), Management Team
From: Sylvia Phillips, Head of BBJ Advisory Team

As Mr. Reed explains, BBJ now needs to decide its role in northern Kyushu. This is no longer an emergency. It is time to pause and plan. Kindly consider the below factors.

Staff Morale

It seems that some of BBJ's staff are really tired. They have been encouraged to volunteer in the affected area, and many have been released from their normal duties as well. It is time to reconsider the situation.

Water Purification Decision

Civil Solutions (CS), a medium-sized engineering company based in Kyushu, has offered to work on the site for free. CS may compete with BBJ in this market over the next few years, so it might be better for BBJ to complete the work itself. BBJ has a good reputation, and it seems that the mayor's office wants BBJ to do the work. BBJ needs to find out fast exactly how big the job is.

Publicity

BBJ's Head of Communications thinks that BBJ should tell the public about its disaster relief efforts. He would like to start TV and press advertising. However, Reed and others think that making a big announcement could be misunderstood as self-serving, with a possible negative effect on public opinion. Some employees and the government authorities may also find such an approach distasteful. This issue must be resolved as part of BBJ's strategy.

Production and Sales

Since the disaster, BBJ has lost 17% of its production and about 26% of its sales across Japan. One reason for this is the lack of trucks for distribution. Also, some raw materials have been difficult to get. From now on, how much production and sales would BBJ be able to sacrifice for disaster relief? Will good CSR (and good PR) create enough long-term compensating value for the business?

Using the company website to collect donations for the relief fund has boosted the company's profile. It is important that the Group Chairman has given his strong support for this work. Share prices remain positive.

There are many more ways that BBJ can contribute to relief efforts in the future. It will take many years, and rebuilding has not even begun yet. The strategy developed now should consider this from a long-term view.

I hope that this additional information is useful.

Sylvia Phillips
Head of BBJ Advisory Team

Change of Direction

The problem for BBJ is no longer simply how to respond to a crisis. Now it really has to plan its own future as a business. It cannot allow competitors to take over its market.

At the same time, in order to increase production and sales, it will have to take some resources out of the relief effort.

As you can see, Reed has two questions to answer at the same time. First, he needs to make a quick decision about the water purification plant. Second, he needs a longer-term strategy that balances BBJ's philanthropy with its business objectives. The purification plant is pressing and makes the question of overall strategy urgent.

After speaking with Sylvia, you decide to invite Fernando Rodriguez to share his professional advice about how to proceed. He is one of the consultants that Pentalemma has used for many years.[1]

To: Pentalemma Inc.
From: Fernando Rodriguez

Dear Sylvia and team,
I use several well-known management tools to help project teams define and achieve their goals. These tools follow the Six Sigma structure of DMAICL (define, measure, analyze, improve, control and leverage). This method has helped many organizations around the world make significant, lasting improvements to their business. In my work, for challenging and complex issues, I always find a way to move forward by using these tools.

I usually like to start by defining the problem in order to ensure that everyone has the same clear understanding of the problem before trying to solve it. Then I start the creative processes. Brainstorming helps to generate new and diverse ideas. Then I use other tools to organize the ideas so that we can focus on finding a solution. There are various kinds of analyses for doing that.

In the end, I use decision-making tools to make the final choices. A common question is how to use action-planning methods to find the way forward by clarifying goals, developing feedback and giving encouragement to reach successful outcomes.

I look forward to working with you during this difficult time to create a good plan for the next steps.

Fernando Rodriguez
Senior Facilitator, FR Consultancy

Five Options

To prepare for the meeting with Rodriguez, you review the five options as they stand in terms of current feedback from BBJ.

Option A—Cessation of Relief Activities

Other companies are stopping their relief activities in Kyushu. Our sales figures show that perhaps BBJ should do the same. Also, some of our staff are getting tired of the volunteer work. There are not enough resources to do relief work and take care of our customers at the same time. Clearly, BBJ should not lose a large part of its market in order to continue with philanthropy.

The water plant in Saiki is a business opportunity. We could offer to complete that work for a fair price. Some say that such a decision may hurt the company's reputation. But if the damage to our image is not too bad, and not too expensive, we could still try to achieve modest growth. BBJ will have to stop at some time, so perhaps we should quit now rather than later.

Option B—Pure CSR

We need to see this situation from a long-term perspective. The worst part of the crisis is over, but we can't simply stop and leave 150,000 people without clean water. BBJ should repair the water plant. The repairs are an engineering challenge, as the plant was totally covered by seawater. However, if we don't do it, someone else will do it.

In the news, we still see images of families in shelters. Companies should wait until the authorities say we can stop the relief work. If we stay, we can expect at least some improvement in our reputation with the Japanese public. That will be a long-term benefit. In the short term, though, our results are likely to be weak anyway.

Full CSR could be very expensive, but we should prevent any perception that we are "abandoning" Kyushu. This is a developing situation and is changing every day. If we stay involved, Japan will remember our support.

Option C—Pause and Plan

We no longer face an emergency situation. We all need some time now to take care of our own business. The big NGOs, the UN and the Japanese authorities have at least stabilized the situation. The private sector should not quit, but we should ask ourselves what we should do next.

This could be a "win-win," because, a dialogue with the other stakeholders could help to develop a more strategic Disaster Plan for Kyushu. It would also give more time for BBJ to set its priorities. This would have benefits for everyone.

Right now, we don't know how much more work there is, and we don't even have a list of priorities. Rather than make hasty decisions, it is better to sit down with other stakeholders and begin to plan properly.

The risk is that some people who are under time pressure to make progress, especially the mayor of Saiki, might disagree if we don't make a quick decision.

Option D—Smarter Relief

BBJ could offer to do the work on the water plant for free. However, an investment partnership with the local authorities is also possible. We should negotiate with them before making a decision.

Meanwhile, it would be reasonable to reduce the number of trucks we are using for relief activities. Others can pick up some of this burden. BBJ needs those trucks in order to recover from the sales slump in Honshu.

The crisis has passed, and a new balance of responsibilities is appropriate now.

That means we should bring most of our staff back to their usual jobs. Perhaps we should also stop the fundraising campaign on our website.

Option E—Partnership-based Commitment

This crisis has given us a lesson in partnership. The government, the NGOs and businesses have found ways to collaborate. Perhaps the best example of this was our work with Telecom Japan, setting up the crisis communications network. We could have more impact if we use this approach.

For example, it would be much easier for BBJ to repair the water plant if the local authorities provide accommodation for our engineering staff. In order to finish what we started, BBJ should complete the work on the plant. This will greatly improve our chances of receiving additional contracts later.

Regarding the relief effort, coordinated planning is the key. This will also provide us with valuable lessons for future crisis management.

Analyzing and Brainstorming

You can see that most members of the BBJ team hope to create a strategy that protects BBJ's interests while helping the communities that have been affected. You and Rodriguez meet with Reed and his team to discuss the strategy. These discussions are often quite serious. On BBJ's side, there is a wide range of opinions on the main issues.

Business Philanthropy

Philanthropy during the crisis has been good for BBJ's image, but sales have fallen due to transportation problems. There are those who argue that BBJ is not a philanthropic organization and that it should get back to business immediately. Other people say that BBJ should continue making contributions to the relief work because it is the right thing to do.

A third group thinks that BBJ should apply its CSR policy because it is part of the community. The problem is how to balance the needs of disaster victims with the needs of other stakeholders, such as shareholders and customers.

Integrated Partnership Planning

BBJ was successful in the early stage because it created partnerships with other companies involved in disaster relief. For example, BBJ trucks were sometimes used to deliver the products of other companies. The partnerships have generated great media coverage and boosted the company's reputation. The people in the local area, as well as the businesses, are beginning to appreciate BBJ as a friend.

Some BBJ managers want to build on these successes. In this way, BBJ's relief efforts could have the most impact. Others do not like working with competitors or potential competitors. They want to stop such partnership work and concentrate on increasing sales in the Honshu area.

Competition in a CSR Setting

Even in a crisis, business is business. The offer made by Civil Solutions to work on the water plant for free is a serious threat to the future business of BBJ. BBJ cannot ignore this threat. Also, the lack of clean water affects hundreds of thousands of people. Should BBJ repair the plant for free? Should it charge for the work like a normal business project, or charge a special low rate? Or can BBJ negotiate a deal that satisfies everyone involved?

Public–Private Collaboration

Some of BBJ's managers want to keep things simple. They do not like being mixed up with government agencies when they manage a project.

However, there are various models for collaborating with the public sector. Reed and some of the other managers see the crisis as an opportunity to use such models. This is a situation where public and private interests are similar. Government agencies seem open to new ideas about cooperation and sharing resources. Reed and the mayor of Saiki, both experienced negotiators, work through the issues one by one with the aim of creating enough value that both sides can feel that they gained more than they sacrificed. By prioritizing the issues and confirming the many points of shared concern, they finally reach a consensus that they can confidently take back to their constituencies.

Having come up with a sustainable and mutually beneficial agreement, BBJ now needs an plan to put it into action. Rodriguez works with your Pentalemma team on strategic planning as outlined in the Project Management Body of Knowledge (see Figure 2.1 below), published by the Project Management Institute in 2017. He talks about the basics of

Strategic vs. Tactical Planning*

Strategic	Tactical
• 3- to 5-year horizon	• Near-term, within one year
• Organization focus	• Department and team focus
• Big picture items	• Project and task level of detail
• Management level	• Operational level
• Often outward perspective	• Often inward perspective
• Organizational scope	• Project scope

Remember that strategic and tactical planning must intersect and integrate.

*Source: Project Management Institute's PMBOK Guide

Fernando Rodriguez Consultancy

Figure 2.1: Depiction of strategic planning according to the PMBOK guide (2017).

planning while showing the team a few slides. He first explains that there are two types of planning—strategic and tactical.

Strategic planning generally covers a period of years. It focuses on the goals of the whole company. Tactical planning is part of the strategic plan. It may be appropriate for projects to be completed within one year, for example. Tactical plans are made for departments and teams within the company.

It is important that strategic planning and tactical planning are coordinated. Reed agrees that BBJ's strategic planning should aim for a sustainable balance between business and philanthropy. A tactical plan will be developed for the water plant. Doing this, Rodriguez explains, requires familiarity with a concept in project management known as the ten critical elements of action. Referring once again to the Project Management Institute's 2017 publication, Project Management Body of Knowledge (see Figure 2.2 below), Rodriguez helps the BBJ team understand how these can be applied to the situation they now face.

The ten parts of project management

Project Management Elements*

• Integration	• Stakeholders
• Scope	• Human Resources
• Time	• Communications
• Cost	• Procurement
• Quality	• Risk

*Source: Project Management Institute's PMBOK Guide

Fernando Rodriguez Consultancy

Figure 2.2: Adapted list of the 10 parts of project management according to the PMBOK guide (2017).

1. Integration
   Project integration starts at the beginning of the project and continues until the end. It requires detailed planning. For example, there should not be too much strain on human resources in order to meet deadlines. At the same time, the project must be integrated with the company's other activities, so that it does not interfere with them.
2. Scope
   Scope management also starts at the very beginning of the project. This is where we develop a clear definition of what must be done. For example, for the water purification plant, will water supply for the plant itself be included in the project? What happens if the city wants to increase the scope of the project? That would probably increase the time and the cost of the project. These things must be confirmed from the beginning.
3. Time
   It is better to have clear forecasts, which include the required times for completing each part of a project. Then there are not so many surprises that could cause delays. We divide activities into smaller, more manageable pieces. This helps everything move faster. It also makes the system more flexible if changes have to be made.

4. Cost
   Cost management means staying within the budget! This requires careful planning, cost estimation, budgeting and control for all the resources used, including human resources. Remember, it costs less to make changes early. When changes are made later in a project, they cost more.
5. Quality
   Successful project management requires clear quality standards and a system for confirming that the standards are met. If not, the project will fail even if it finishes below budget and ahead of time. Someone must be directly responsible for every aspect of quality.
6. Stakeholder participation
   Clients sometimes ask to increase the scope of the project. Other stakeholders, such as communities affected by the project, may have other requests. The project team should be in regular communication with the relevant stakeholders, providing information and addressing their concerns.
7. Human resources
   Avoid over-using or under-using staff. A project is an opportunity to bring in new human resources and develop them. Experienced people can build their leadership skills, while younger employees can gain experience.
8. Communications
   A good communications system informs all project members of their responsibilities. They need to understand each other and the status of the project. Images can be used to show facts, such as how much of the budget has been used, or where gaps are in the workflow. Finally, an effective communications system helps everyone understand the next steps and what they must do.
9. Procurement
   The project can only be completed on time if all the resources are in the right place at the right time. All procurement decisions must consider factors such as delivery times and storage (in case of weather damage or theft).

10. Risk

Areas of risk must be found in the early stage of a project so that they can be controlled. The process should have no mistakes. It costs less to prevent mistakes than to fix them. There must also be plans for a worst-case scenario.

Project Management Tools

BBJ uses special software, which they developed themselves. It provides efficient management and presentation of project data. There are many similar software tools available today. They identify project details such as how much time is needed to finish each part of the project.

With this software, skilled project managers can often see how to shorten the time spent on projects. Also, if a change must be made to one part of the project, they can see the effect of that change on other parts of the project.

Rodriguez worked together with Reed and his managers to confirm that they are using their tools correctly and that the ten key elements of project management were all included.

Conclusion

Pentalemma challenges you to develop solutions that benefit all stakeholders. To do this, you have to understand the different perspectives among the stakeholders and help them develop sustainable agreements. BBJ still has stakeholders who like pure philanthropy, while others think BBJ's profit is the highest priority. They see philanthropy as an unnecessary expense. Your challenge is to support BBJ so they can make the best decision for themselves and their stakeholders. There are five options on the table—how will you advise your client?

Once again, let us fast-forward to consider five hypothetical sets of consequences.

Initial Impact on Reputation

Option A—Cessation of Relief Activities

Kelvin Reed in Row with "Snouts-in-the-Trough" Journalist

There was a stand-up row when Australian journalist Phil Johnston attended a BBJ press conference yesterday. The meeting, arranged to publicize BBJ's plans for a new line of baby foods, degenerated when the Australian berated BBJ for its withdrawal from the Kyushu relief effort. Mr. Johnston has already published a controversial piece entitled "Snouts in the Trough," in which he alleged that BBJ's only interest was in making money out of Kyushu. Mr. Reed described Johnston as a "disgrace to journalism."

Option B—Pure CSR

Residents to Wait Another Two Weeks for Water

Finally, BBJ has stepped up to the plate and agreed to refurbish the damaged water plant at Saiki. Local people have been without clean tap water in the area for two weeks now—and it could well be another two weeks before the situation is restored. BBJ has explained that repairing the plant is an engineering challenge, as it was totally engulfed by seawater. Nonetheless, there had been a two-day impasse when the Mayor sought support for the work to be done pro bono. "I don't have the funds to release for this just now, but the work can't wait," she insisted. BBJ said it was happy to do the work for free, but had been forced to delay other engineering work to the south of Kyushu.

Option C—Pause and Plan

Saiki Water Plant: Mayor Will Use "Her Bare Hands"

The Mayor of Saiki has invited residents to help clear the rubble and silt from the town's flooded water plant "with their bare hands." In an emotional appeal on Thursday's evening news, she said that people couldn't wait any longer while the aid agencies and others twiddled their thumbs. Mr. Reed, who is at the center of the storm, appealed for cool heads. "Digging around in the mud might make good television but won't make the slightest difference to the engineering problem. We need a chance to think this through."

Option D—Smarter Relief

Mayor Signs Water Deal with BBJ

Saiki's Mayor today signed the first major redevelopment contract for her town with BBJ. The MNC will rebuild the town's damaged water plant, restoring clean water to over 150,000 citizens. In what is seen as a template for other work, BBJ has agreed to do the work "at cost," which means the Mayor's office will get the work done at the knock-down price of $1.5m. "It's a gesture," said Mr. Reed. "We've been able to do a bit of emergency work here, and now this enables us to make a small contribution to the town's future as well." The restored plant is expected to reopen in two weeks.

Option E—Partner with Other Stakeholders

Special UN Study on Kyushu Relief Work Announced

The United Nations Foundation is to commission a special report on the teamwork which has been pioneered recently in northern Kyushu. UN envoy Gianni Franchitti cited the partnerships put in place by the private sector for particular praise. "What we have seen here is an interest in getting the best from the different players involved. So what Oxfam is good at doing, it has done; the same is true of KZ Foods and others. It has been less of a scramble and more of a coordinated effort, with very little jealousy and rivalry on the ground. This is very new. And we are sure it bodes well for the rest of the work that needs to be done here, but also, for other crises around the world. That's why we want to study it and to document it."

Later Impact on Reputation

Option A—Cessation of Relief Activities

Media Reaction

Campaigning Australian journalist Phil Johnston's piece on the MNCs in Kyushu has done extensive damage. The story, which also ran in Japan, the UK and the USA, was entitled "Snouts in the Trough." Unfair though it was, it highlighted the withdrawal of firms like BBJ from direct aid, only to return to the area to bid for commercial projects such as the Saiki water plant. Johnston's rather shrill call for a product boycott seems to have had little impact to date, but the reputational damage is undeniable.

Option B—Pure CSR

Media Reaction

Ultimately, the provision of a revamped water plant just wasn't an interesting enough story for most of the media. The continuation of "Click Brighter Brand" works well for BBJ (and for Telecom Japan), but the news has now shifted focus to other aspects of the quake story.

Option C—Pause and Plan

Media Reaction

The Mayor of Saiki was on prime-time TV on Wednesday night slating BBJ and others for lack of activity. Mr. Reed's defense was adequate, but the mayor's charge was difficult to refute in light of continuing delay with the water plant. The coverage of families still using bottled water after weeks did not help either. BBJ's public stock is definitely down as a result of this.

Option D—Smarter Relief

Media Reaction

BBJ received some plaudits for its partnership deal on the Saiki water plant, with the Mayor going out of her way to describe it as "the way forward." Phil Johnston, writing in the New Statesman described it as "vulture capitalism," but this line did not gain any traction elsewhere.

Option E—Partner with Other Stakeholders

Media Reaction

"Brighter Brand Leads the Way in Kyushu!" ran the front page of Britain's Economist last week, a line quickly picked up in Japan and elsewhere around the world. BBJ's coup in bringing KZ Foods back to the relief work was widely heralded, along with the creative partnerships it has built with Oxfam and the civic authorities. It seems the media has finally "got it."

General Impact on Firm Value

Option A—Cessation of Relief Activities: Return to Commercial Priorities

COMPANY REPUTATION: BBJ's decision to halt its relief efforts after only ten days caused a massive dent in the company's reputation. At a time when so much help was needed, cutting and running did not sit well with the locals or the media. Also frowned upon was BBJ's offer to clean the Saiki plant at their standard commercial rate, especially when its competitor Civil Solutions offered to do the job for free. This all equates to a negative $5,000,000 net value for Brighter Brand Japan.

SALES: The hit to the company's reputation was reflected in its sales performance. There was zero growth from last year due to the negative PR campaigns and news of BBJ's offer to clean up the Saiki plant for a fee. Although BBJ eventually secured the contract at a value of $2,000,000, the damage was done—and this hardly dented the overall trend. Needless to say, this posture has caused measurable harm to Brighter Brand Japan's growth plan for the year, and we estimate this to reduce sales to a net value of only $30,000,000.

RISK MANAGEMENT: BBJ scored a huge own goal here in terms of its PR operation. In order to offset a further slide in sales, it has become necessary to commission a special advertising campaign, which will run from the autumn. Alongside the more conventional risk management measures that have been introduced in the aftermath of the quake, this has impacted on the overall value of risk management. The net value of risk management is currently assessed at $2,000,000.

SAVINGS IN SUPPLY COSTS: Pulling the lorries out of Kyushu may have helped the transport situation, but crucial raw materials remain at a premium in this part of Japan. Aluminum prices have shot up by nearly 20%, for example. In addition, BBJ plants located nearest the disaster remain at a 50% running capacity; the roads are full of government transport, and the goods cannot get through. All told, BBJ did not meet its expected savings and only achieved a net value of $10,000,000.

STAFF MOTIVATION: After working tirelessly for ten days with only minor breaks, staff morale and motivation began to take a dive. Even those who were relieved from their normal jobs had to contend with fatigue. This tendency was only exacerbated by what has been portrayed as a "cut and run" strategy by staff representatives. The provision of some sensible rostering and the continued support of those staff who remain keen to continue with the relief work might have been a smarter approach here. There have been a number of resignations, and sickness levels are the highest ever reported by BBJ. The value of staff motivation has plummeted to only $4,000,000.

Option B—Pure CSR: Staying Committed to Those in Need

COMPANY REPUTATION: Brighter Brand Japan enjoyed positive publicity for its decision to continue full-scale relief efforts in the

Kyushu region. This increase was significant. Conversely, investors realize all corporate social responsibility comes at a price and will be reflected in the bottom line. Even with improved long-term prospects for future market share, the stock price has dipped since the beginning of the year. All told, the positives of staying to help equaled the negatives of declining market value. Therefore, there is no net value added to the company's reputation per se.

SALES: Sales have been suffering. Continued CSR support meant staff extractions and consequent loss of production. A more balanced commitment, perhaps in "smart" partnerships with others working in the area, might have resulted in less disruption. BBJ's response to the crisis is recognized by the Japanese public, but it has not always translated into extra sales—particularly when the goods aren't on the shelves. This created a net value of sales of a modest $45,000,000.

RISK MANAGEMENT: Considerable new risk mitigation measures have been necessary in the aftermath of the recent earthquake, and these have already built new costs into BBJ's model. Unfortunately, the level of commitment still being made toward relief efforts has made it particularly difficult to assess BBJ's ability to recover its business as usual. Some market share may be gone forever unless special remedial measures are introduced. Exactly how this transpires will depend on firmer figures further into the year, but it is clear that BBJ will need to invest in order to protect its position. Risk management factors, therefore, netted only $5,000,000 at the latest assessment.

SAVINGS IN SUPPLY COSTS: With most of our shipping resources being utilized (and not reimbursed) to aid refugees from this earthquake, BBJ has so far been unable to realize its expected savings in supply costs. The lorries that were taken out of commission caused our supply chain to break down, and the ferries that we generally use were also dedicated to disaster relief. Even a modest redeployment of the transport fleet could have avoided this. The problem has been the 100% CSR emphasis, which was agreed by the Board. BBJ's estimated net value of savings in their supply chain was negative $5,000,000.

STAFF MOTIVATION: The morale of the staff now seems to be subsiding, after an initial surge of support. We see this in stress-related illness, requests to return to normal working and an uptick in turnover. For the workers, each day brought more of the same and kept them farther away from performing their real jobs. Efforts to replace them were of no avail, and they continued to work outside of their jobs for much of the time. Earthquake fatigue seems to be setting in. Estimates put the net value of staff motivation at negative $5,000,000.

Option C—Pause and Plan: Develop a Multilateral Plan

COMPANY REPUTATION: We saw a divergence of opinion about BBJ in light of the recent decision to develop a multilateral plan with the UN and other agencies. Stock market confidence remains solid, with shares increasing in value by nearly 2% in the week since this was announced. Market surveys continue to show full consumer support for BBJ, both in Japan and across the Asia-Pacific area. All this against a backdrop of some furious press coverage, which could have damaged the company badly. The Mayor of Saiki has been posturing about the delay with the water plant, but somehow such views did not gain traction. Reputation posted a healthy net value of $10,000,000.

SALES: Sales were able to maintain levels close to this time last year but naturally dipped slightly as it took time for operations to resume their normal course. The pause in BBJ's activity within the disaster zone has freed up some assets and helped production to recover. The value added was calculated to be $60,000,000.

RISK MANAGEMENT: The let-up in the pace of relief work provided breathing space to the company which was put to good effect. Some of the commercial ground lost has now been recovered, repairs to plant undertaken and thought given to how many resources might be required for future efforts. All of this makes both long- and short-term risk planning easier—and cheaper, in that we have a clearer understanding of where we might be headed. Risk is currently assessed at a net value of $1,000,000.

SAVINGS IN SUPPLY COSTS: With business returned to semi-normal operating standards, BBJ was able to move ahead with plans to secure cost savings in the supply chain. While this was dampened during the initial days and weeks after the disaster, the company was able to regain control of almost all of its lorries and ferries for purely operational purposes. BBJ estimates these savings at a net value of $20,000,000. Certainly, this pause has enabled BBJ to realize some big savings.

STAFF MOTIVATION: Returning to their normal jobs was one of the major issues that concerned workers, according to this month's staff survey. The company's current posture has allowed many to return to their day job and in some cases, to take a holiday. Reports of burnout have declined sharply. This shift toward a normal pace of life has helped with morale and motivation. Staff motivation is now valued at $8,000,000.

Option D—Smarter Relief: A Measured Approach

COMPANY REPUTATION: BBJ enjoyed some kudos in offering to repair the Saiki water plant at cost, and the Mayor was gracious enough to sing the company's praises. The Japanese public remains well disposed toward BBJ, as evidenced by our surveys as well as product demand. Shareholders remain loyal—with an essentially stable price. The net value of this approach was $15,000,000.

SALES: Sales dipped in Honshu and Kyushu. Despite this, demand is high and deliveries are getting through to the rest of the country. The outlook is definitely rosy as long as production and delivery can meet demand. The company expects sales to return to within normal parameters in the coming weeks. BBJ estimates a net value to sales of $70,000,000.

RISK MANAGEMENT: Reputational risk has clearly been well managed here, with a well-balanced and well-considered plan. The reallocation of much of the transport fleet addresses the risk that BBJ will lose market share on a permanent basis, and in overall terms, the company's risk profile is now well understood and appropriately managed. The company assumes a net value-added of $7,500,000.

SAVINGS IN SUPPLY COSTS: Savings were not realized by BBJ in the current period due to its use of fleet and lorries to aid in the disaster efforts. BBJ had projected a positive net value added of over $20,000,000, but without being able to benefit from purchasing power or economies of scale, this figure was only $5,000,000. However, we now expect this situation to improve rapidly and for future projected savings to come in on profile.

STAFF MOTIVATION: Realism and balance have been introduced into BBJ's continuing involvement with the Kyushu earthquake zone. Staff surveys confirm a broad acceptance of this stance and a willingness to deliver against this more modest agenda. Attendance and productivity have been normal in recent weeks. Staff motivation was up over the last report and BBJ estimates its net value added to be $9,000,000.

Option E—Partner with Other Stakeholders: Full Integration

COMPANY REPUTATION: All parties involved—UN, NGOs, government and even competitors—were impressed with BBJ's offer to perform the work needed to repair the Saiki water plant at cost. They were also praised for being the first to set up disaster assistance on their website, including matching funds donated. Bringing KZ Foods back into the disaster zone was seen as something of a coup and the piece in the "Economist" played brilliantly with the markets. BBJ's advocacy of mature partnerships has contributed to what is an all-time high for its reputation. All in all, this generated $25,000,000 in the value added to our reputation, as future relationships will allow for long-term growth.

SALES: By partnering with KZ Foods to optimize transport assets, BBJ managed to improve on deliveries elsewhere in Japan. BBJ thereby successfully protected sales volume for the year, which is now slightly up. Future partnerships that are being discussed among the parties involved with disaster relief are sure to benefit the company in the long-term as well. BBJ believes this will add $140,000,000 in sales value.

RISK MANAGEMENT: The development of strategic partnerships with companies such as KZ Foods and Telecom Japan means risk sharing as much as anything else. This could work for the long term as well as the short, with early discussions between BBJ and Telecom Japan already bearing fruit. The current risk landscape is now well understood and controlled. The risk for the year ahead may well be better managed as a result of this initiative. Risk management is currently valued at a healthy $15,000,000.

SAVINGS IN SUPPLY COSTS: Supply difficulties continue in Japan, as much of the infrastructure has yet to be repaired, and many vital goods remain in short supply. Some synergies have been achieved, particularly by working jointly with KZ Foods. As a result, modest savings of $10,000,000 have been achieved.

STAFF MOTIVATION: Creative planning with partners on the ground has meant a greatly reduced burden on our own staff. Normal working patterns are beginning to return, and stress levels are declining. Yet there is still plenty of commitment from those involved. This seems to be a healthy balance: productivity and sickness are close to normal levels. Support for BBJ's corporate response is at an all-time high, as are applications to work at the company. Staffing is valued at $15,000,000 currently—highly impressive under the circumstances.

Reflecting on Chapter Two

The Scenario in Context

This scenario opened only ten days after the tsunami and nuclear accident which had devastated the northern coast of Kyushu. The pace of decision-making and unfolding events was, therefore, more intense than in earlier Pentalemma scenarios, with the sense that the wrong call might quickly prove disastrous for BBJ. Despite this, it was clear that the immediate emergency had been addressed and that what might be called "the proper authorities" were now in control. A good time then, for BBJ to reconsider what it should be doing in the region. Normal business had clearly been suffering, and so there was an argument for ending disaster relief, as other companies had done. By the same token, there was now the possibility of reconfiguring BBJ's support, with a new emphasis on rebuilding.

Dilemmas and Challenges

The tactical question that the BBJ management team needed to address most urgently was whether to rebuild the water plant. There was a call from the local government to contribute to this project, although this could hardly be regarded as an "emergency" when compared to the need to rush food and blankets into the area that had been so critical only days before. Such work was normally a bread and butter commercial activity for BBJ. Why not treat it as such?

This problem was compounded by the involvement of a company called Civil Solutions (CS). CS, a direct competitor to BBJ, had offered to do the work for nothing. Should BBJ enter into some kind of bizarre "bidding war" with CS, along the lines of which firm could offer the best CSR deal? Or were such blatant commercial considerations inappropriate under such circumstances? The backdrop remained whether or not BBJ should continue its relief work at all—sales had slumped in other parts of Japan. Specifically, the lorries which to date had been running supplies into the stricken zone were needed to resupply retail outlets. Might a reconfiguration of BBJ's posture, with a continued presence in the area, but based on a defined reconstruction project, be the way forward? How would BBJ's reputation respond to these various options?

Arguably, the strategic decision as to whether or not to remain engaged was contingent on how things actually panned out on the ground. It could not be considered in academic isolation. This is a challenge commonly met in business.

If BBJ decided to remain involved, managers would then need to consider how to do so. There was a myriad of factors to look at—including an exhausted workforce, with accompanying employer responsibilities, the possibilities for further "disaster partnership" with companies such as KZ Foods, and the public relations repercussions. Should BBJ offer to repair the plant at cost, for free or at a profit? Did a commitment of this sort imply pulling back from the earlier work in direct disaster relief? Was Pentalemma able to contribute positively to BBJ developing a win-win situation.

Note

1. For the past 15 years, Fernando Rodriguez has trained management professionals in the latest tools and techniques, building for himself a reputation for helping companies develop lasting improvements. His approach begins with building consensus on the definition of the problem to be solved. From there, he employs several methods of gathering ideas, such as brainstorming and mind-mapping, and then harvesting potential solutions by using an affinity diagram or similar tool. In short, Fernando works with managers to clarify their goals, and agree on the changes they need to make to achieve them. In his words, "I simply help managers become dynamic agents of change themselves."

Further Reading

Elkington, J. (2002). *Cannibals with forks: the triple bottom line of 21st century business.* Oxford: Capstone.

Freeman, R. E., Harrison, J. S., & Wicks, A. C. (2007). *Managing for stakeholders: survival, reputation, and success.* New Haven, CT: Yale Univ. Press.

Galtung, J. (2012). *Peace by peaceful means peace and conflict, development and civilization.* Oslo: Prio.

Kahane, A. (2012). *Transformative scenario planning working together to change the future.* San Francisco, CA: Berrett-Koehler.

Nelson, J. (2000). *The business of peace: the private sector as a partner in conflict prevention and resolution*. London: International Alert.

Project Management Institute. (2017). *A guide to the project management body of knowledge: (Pmbok guide)*. Newtown Square, PA, USA.

Chapter Three

Logical Thinking

You have been advising BBJ since the huge earthquake and tsunami hit the northern coast of Kyushu four months ago. BBJ remains engaged in disaster relief, but the focus is changing toward the longer term. The Japanese government is considering major issues, such as the nature of a future economic framework for the area, together with supporting infrastructure, housing, etc.

There will be a lot of work to do, entailing major public investment and possible opportunities for BBJ. The question of whether BBJ wishes to be involved or not remains. At this stage, any participation would be for the long haul, with the company invited to take a serious stake in the future of this community.

Your own follow-up research reveals several important factors. A central concern for BBJ in this situation is its need for a new polythene plant to serve its Japanese operation. Had the Kyushu disaster not taken place, it seems likely that the plant would have been built on the main island of Honshu, closer to BBJ's largest Japanese markets. Now, though, there is talk of large government subsidy for new industry in Kyushu, alongside the development of brand-new infrastructure.

Therefore, the cost-benefit analysis for a new polythene plant will have changed. BBJ managers will need to think about these new factors as well as the broader principle of community engagement. It is entirely possible that building in Honshu remains the most commercially attractive option. Additionally, there may be a whole raft of other profitable opportunities for BBJ as the area rebuilds, irrespective of whether or not it locates its new plastics plant there.

This picture is made more challenging by some fairly heated local opinions. Many residents hanker for a revived fishing industry. Yet even before the quake, this sector was in decline due to over-fishing. The environmentalists, in the form of "Green Action" oppose this, as well as the idea of a plastics plant. They have a radical agenda for a local "green economy." The politics, therefore, seem toxic. As such, it may simply be better for BBJ to disengage. In doing so, they might make a big final gesture, such as rebuilding a local school or sports facility.

A set of difficult decisions requires a discussion of several topics:

1. Investment location decisions for major business projects
2. The adaptability of the "green agenda"
3. The role of business in society

Subjects Covered

The many issues raised in the scenario include fishing and the environment; rail links; housing; sports and education facilities; supplier and customer industries; portable development templates and government subsidy.

Scenario Design

The idea with this situation is to introduce some of the issues likely to arise as an area emerges from a crisis situation. It is at such times that some of the broadest and most important planning decisions need to be taken. These can be looked at in isolation, or more usefully, in an integrated manner, with all parties involved. As is common in Pentalemma, the scenario was designed so that a realistic and immediate "micro" challenge for the company—in this case, the need to build a new polythene plant, had obvious relevance to the broader (or "macro") dilemmas facing the local area.

Alternatively, looked at purely from BBJ's perspective, the macro issue might be whether or not to engage with redevelopment in Kyushu. The micro issue could be whether to build the new plant there. The scenario also seeks to underline how strongly held views about the environment, and traditional lifestyles can feature in these debates. Such sentiments are just as important in the "developed" economies as they are elsewhere. Both the environmental and fishing lobbies in modern Japan are vocal and influential. But it would be as easy to find such conflict in (for example) the recent oil shale "fracking" debate in parts of western Europe. Similarly, for all of their strident expression, there is usually room for compromise with such groups.

Finally, there is an ambition to stress the symbiotic relationship between businesses and the societies they serve. Kelvin Reed may have made his point about Japanese society from a moral perspective, but there are solid commercial reasons to regard this relationship as integral. The future development of Kyushu in this scenario is replete with potential synergies, which can be of benefit to all parties. These range from the ability to reconcile the differing views, to securing complementary industries and infrastructure, such as housing and new transport links.

In this final scenario, Pentalemma advisors and BBJ managers must unpick a tangle of seemingly competing interests emanating from environmental, business and social concerns. By looking for those risks that can be turned into opportunities, they should be able to identify positive-sum solutions, whereby all parties benefit. In real-world crises this is difficult and not always guaranteed, yet it makes good business sense—and is worth striving for.

Questions to Consider

1. Is Kelvin Reed right to argue that BBJ is "part of Japanese society"?
2. Having opposed the building of a polythene plant and any resumption of fishing, might Green Action moderate its stance? How would such a shift be brought about?
3. Apart from it being closer to BBJ's main markets, what other considerations might support a decision to locate the new polythene plant in Honshu?

Significant risks related to continuing involvement in northern Kyushu include:

Political: A new plant in Kyushu would satisfy some segments of the population, such as those who have lost their homes and jobs. But this could aggravate other groups like Green Action. BBJ needs to take both directions of reaction into account before making a final decision.

Environmental: Although perceived by many as a radical environmental group, Green Action's concerns around the increased air pollution that would result if a new plastic packaging plant were built ought to be examined carefully.

Demographic: Five million people live in the area where transportation infrastructure has been seriously affected by the earthquake. This is a significant factor to take into account when deciding whether BBJ should remain engaged in the community.

Economic: As many of the transport links have been destroyed, the cost of shipping and supplies for building a new plant has skyrocketed.

Social: Although helping repair the damaged fishing industry in the area would require a long-term commitment with uncertain media exposure, many of the residents and the Japan Fisheries Association remain hopeful that BBJ will lend support.

Kyushu Recovery: Future Involvement

BBJ Head of Operations Kelvin Reed has cc'd you in an email to your team leader Sylvia Phillips.

To: Sylvia Phillips, Head of Pentalemma Advisory Team
From: Kelvin Reed, Head of Operations, Japan
cc: Duncan Bruce, Group CEO.
Re: Future involvement by BBJ in the rebuilding program, Kyushu.
Sylvia,

It has been four months since the earthquake, tsunami and nuclear accident in Kyushu. This is a good time for us to review our role. We have earned an excellent reputation for our community support. We responded quickly, immediately following the earthquake. We sent goods, raised funds and encouraged our staff members to volunteer. Some of these activities continue. Sales are now only 1.5% lower than before the disaster.

A huge rebuilding plan, including infrastructure, new industry, housing and agriculture, is under discussion. There will be opportunities here, but there are also risks.

We could return to normal business and look for opportunities created by the rebuilding plan. Yet BBJ is not only a business. These days, it is also part of Japanese society.

For the medium term, BBJ could continue raising funds and offering some direct support, such as sending staff or providing technical and business advice.

On the other hand, we could set an example by building our new plastic packaging plant in Kyushu. Damage to the fishing industry here has cost many jobs. The community needs economic activity. At the same time, Green Action, the environmental group, is touchy about the overproduction of plastics and will worry that the plant would cause air pollution.

I would appreciate some advice and discussion on these issues.
Kelvin

Sylvia Phillips has informed Reed that you are the person directly responsible for this part of the project. She has also shared with you the following report.

PENTALEMMA INC.: BOARD BRIEFING PAPER
Client: Brighter Brand Japan
Region: Kyushu, Japan
Topic: Japan disaster—Long-term engagement
Author: Raj Patel, Head of Risk Management, Pentalemma Inc., Geneva
Following a period of urgent reactions to the disaster, northern Kyushu is now starting long- term rebuilding. BBJ must decide how to balance its business activities and philanthropic activities in the region.

There are opportunities, and there are threats, as mentioned in Kelvin Reed's summation. The plastics plant option would need much further study, but various factors should be considered from the beginning.

The Fishing Lobby

The fishing industry in this area has been severely damaged. However, many people believe it can be saved. There are strong opinions about future employment and the Kyushu economy. The idea of building a plastics plant would surely cause a lot of debate. It might be better for BBJ to avoid that situation. BBJ can contribute in other ways that do not have as many risks.

Traditional Company Support

BBJ is a big company with many activities. Therefore, it can find various ways to get involved in the area. At the same time, it should continue its philanthropy. Otherwise, some people may think that BBJ is profiting from the community's suffering.

Transport, Infrastructure and Labor

It would cost a lot for BBJ to build its plastics plant in this area. Costs, both for building and operating the plant will be high because of the current damage to railways and roads. Also, will demands by Green Action for a "green" environment create unreasonable costs? On the other hand, labor costs in this area will be quite low compared to other areas of Japan.

Simple Sponsorship

A plan for a school building was recently announced, and the government seems open to sponsorship ideas. By helping with the construction of schools, BBJ could make a safe, long-term investment in the community.

BBJ must first decide whether it wishes to remain active in the area. If it does, there is a wide range of options. Each option has costs, benefits and risks. These should be carefully assessed. Pentalemma is fully committed to assisting with this process.

Raj Patel,
Head of Risk Management

As Pentalemma advisor on the spot, your challenge is to help BBJ create a strategy that protects its own interests while also helping the communities that have been affected. Working with Reed's research team, you have developed five options for everyone to consider.

Option A—Business as Usual

Protest could break out if BBJ decided to build a new plastics plant here. We should stop our support work, even though we would probably lose the business partnerships we have built since the disaster.

There will be opportunities for business, including projects supported financially by the government. Without such strategic partnerships, it will be difficult to get new contracts.

If we leave Kyushu, the community can plan its future. Social and environmental issues are two very different matters. Business should not decide cultural and social priorities for the community.

Option B—Major Final Action

Debates about the future of this area are complex and difficult. Therefore, it would be wise for business to stay out of the way. However, it would be appropriate to make a donation. We could, for example, ask publicly for large companies to provide the ¥72,000,000 needed for the Kyushu Fund for Children. In addition, BBJ should fund a significant project of its own, such as building a school, without getting involved in community affairs.

Because people are listening to Green Action, there is probably no chance to build a plastics plant here. For the future, BBJ can build up its business in the new economy for the region. Industry should help the people of northern Kyushu with one final donation so that they can then determine their own future.

Option C—Let the Community Find the Solution

There are difficult environmental and social issues in northern Kyushu. These issues must, first of all, be carefully considered by the people who are most affected. Does the community want to try a "green" economy? Do people believe it is a good idea to save the fisheries? Or do they want to focus more on other business? It is entirely normal for companies with an investment in the area to contribute to this debate.

The plastics plant will have to be built somewhere. Partnerships and government subsidies would make it possible to build the plant in Kyushu. NGOs, such as Green Action and Oxfam, as well as the Japanese Government and other stakeholders, all have something to say. But finally, it is the community's decision. Only after that becomes clear, business can consider what role it should play.

Option D—Practical Compromises

We have already seen some excellent examples of sharing the workload in response to the Kyushu earthquake. That sets a standard for the future. We can see practical compromises for most of the issues that are currently causing concern. For example, in a meeting with Green Action leaders a few days ago, a rough plan for a "clean" plastics plant was agreed upon. Electric vehicles would move the product to a new storage area.

BBJ has agreed to work with the local government on a housing plan for BBJ workers. We have also requested that the government provide a health center. The future can be built piece by piece, with business playing an important and balanced part. If business is involved, progress can be made fast in a practical way.

A new plastics plant requires that the new railway link is ready on time. Specialist experts estimate only a 50-50 chance of the railway being built on time. We need to understand how these risks will influence the cost of the project.

Option E—Development Council

BBJ has an opportunity to build upon the positive partnership work that has been completed in northern Kyushu in recent months. NGOs, government, the community and businesses have created partnerships that can adapt to change. Now, for building new schools, BBJ could partner with Telecom Japan, which can deliver a modern IT infrastructure at a special rate.

The suggestion is to establish a Development Council with members from various backgrounds. Local people would have most of the stock in the Council.

Investment from the local government would be expected. Environmental issues would be handled according to local opinion. In this way, the community would decide industry's role in their area. This idea was inspired by creative and positive partnerships we have already seen in northern Kyushu.

Logical Thinking Workshop

After conferring with Patel and Phillips, you decide to ask Fernando Rodriguez to come back and help the team decide the next steps. Pentalemma has been working with Rodriguez for many years and has always benefited from his support. He has agreed to lead a workshop for Reed and his team of executives and managers.

Rodriguez starts by explaining that the key to solving a problem is often making a connection between two ideas that have no obvious connection. "Everybody experiences this. It seems like good luck, but we can make it much more than that. There are tools we can use to make it easier to connect ideas," he says. He has the keen attention of all present. "We use Logical Thinking to organize the many ideas that we get from brainstorming so that the best ideas can become effective actions."

He has a pack of colored post-it notes and explains that, in brainstorming, everyone writes their ideas on these sticky papers. Each sheet has one idea written on it. He tells the group that today they will brainstorm about the question: "How to respond to the disaster?"

In the first ten minutes, they produce these ideas:

Rodriguez posts the papers on a large board in front of the group. "That's good," he says. "We have ten ideas. Now we want to make connections between these ideas. We call this the Silent Sort."

He tells each member of the group to stand up and move to the board, where they start picking up post-its and placing them next to other post-its wherever they think there is a related idea. Everyone joins in organizing the ideas. Without saying anything, they sort the random list into groups of similar points.

It is easy to see why this method is so useful. It really encourages everyone to participate and be creative. It supports breakthrough thinking. The steps are very simple. It also moves fast because, at this stage, there is no debate or discussion.

How to respond to the disaster?

Send volunteers	Raise funds
Follow others	Get more data
Ask HQ	Focus on business
Partner with govt	Special campaign
Send supplies	Get staff opinions

Figure 3.1: List of ideas generated by brainstorming in silence.

Now Rodriguez tells the group it is time to give a name to each group of points. That's when the discussion starts. They agree, for example, that "CSR" is the main idea that connects "Sending volunteers," "Raising funds," "Sending supplies" and "Partnering with government."

The BBJ team can now step back and look at all the points organized into categories with other similar points. Rodriguez tells them that they can now discuss which categories are more important or more urgent. They begin to prioritize the categories.

You can see that this is effective because all the communication barriers are gone, and everyone understands the ideas very well.

> "Now that you have come this far," Rodriguez continues, "You can start planning what to do about a certain set of issues. You make a project team for each category."

In only about half an hour, a group of people with various ideas is already able to move into action. The main activity was to find the connections between the initial ideas.

Logical Thinking

BBJ managers are satisfied with the results of the workshop. They say they were surprised to see that, after the "silent sort," they really began talking to each other. They agree to try it again a couple of times to master the process.

Rodriguez watches how they work and provides reminders from time to time, such as: "Remember, the sorting is silent — no discussion yet about which ideas are the best ones."

By the end of the Logical Thinking workshop, Reed and the BBJ team are confident that they can use this method to answer the questions now facing BBJ.

The Logical Thinking question might be: For BBJ in Kyushu, what is the best combination of business and philanthropy?

Reed asks you and Rodriguez to join meetings where the five options can be discussed. They want to think about their impact on the company's reputation, sales performance, risk management, savings in supply costs, and staff motivation.

Initial Impact on Reputation

Option A—Business as Usual

Business Guru Slates BBJ for "Paucity of Vision."

On a brief visit to Japan, American business guru Paul Franklin has slated BBJ and others for their recent decisions to return to business as usual so soon after the Kyushu earthquake. Mr. Franklin, who's bestseller "It's not Just the Bottom Line" has just been published in Japanese, has taken a keen interest in the involvement of business in the disaster relief. He told TV5: "The work BBJ and others did in Kyushu was ground-breaking; now they are simply pulling out rather than recognizing that they have come up with a new model—it's a different kind of business-as-usual, but they can't see it."

Option B—A Major Donation as A Final Gesture

Reed Signs a Check for ¥20,000,000 within Five Minutes of Mayor's Request

In an astonishing act of corporate altruism, Kelvin Reed of BBJ reached into his funds yesterday in response to a plea from a local mayor. Within minutes of the Mayor of Saiki's TV request to top up her Children's Fund, BBJ had come up with ¥20m and was leading the drive to secure the balance of the monies needed to rebuild local schools. By the end of the day, the Mayor had her funds. "It's a farewell gesture from us," said Mr. Reed. "We feel BBJ has done all it can for the area and we leave our donation for the schools as we move back to our day job. We wish the people of the area all the best." Indeed, northern Kyushu has a big debt to BBJ: The company's lorries were some of the first on the scene when the quake hit, and it has been actively working in the area for over four months. This seemed an appropriate closure.

Option C—Let the Locals Figure it Out

BBJ Dithers while KZ Foods Cleans Up

KZ Foods has signed a major building contract in the Saiki area for the construction of an organic food processing plant. In a move described as "an act of faith in this area," KZ Foods will ultimately employ up to 150 people in its new plant. Meanwhile, others stand accused of timidity as local communities begin to think about reconstruction. Asked why BBJ had no plans for the area, its operations chief explained: "We need to see what kind of future local people want before we can determine whether we can become a part of that."

Option D—A Plastics Plant Delivered Through Pragmatic Compromises

Locals Remain Wary of "Green Polythene" Plant

A rare alliance between big business and Japan's radical environmental lobby was struck yesterday. Green Action's Abe Kiyoshi is seen shaking hands with Kelvin Reed, while the Mayor of Saiki looks on. The reason? A state-subsidized polythene plant, using the latest "green" technology, will be built at Saiki, representing one of the first major reconstruction projects. It was only after days of debate that Green Action leaders were able to announce that they could support the plan, which has a neutral carbon footprint. But is this a one-off? The local residents' association has claimed that more is needed in terms of health and housing. They have also pointed out that Saiki is a fishing town, not a plastics one. Progress, yes, but plenty more debate ahead.

Option E—A Development Council

"Silver Lining" Attracts $3bn in New Investment within a Week

The Silver Lining Development plan published only a week ago has attracted a potential $3bn in new investment since its launch, it was revealed today. Former fisherman Shinji Motegi, who chairs the Development Council that put the plan together, is cautious. "What these people need to understand," he says, "is that we don't want just any old investment. It needs to fit with what we have decided is the vision for our community. Is it green? Does it dovetail with any of the businesses we already have signed up? What does it put back for local people?" Motegi is delighted at the response he has had so far, which builds on already firmed-up projects involving companies such as BBJ. He is very clear, though, that aspiring investors need to "get into town and talk to people." But he thinks his plan will work: "If a third of these bids prove good for us, that is still a billion dollars in new money, plus the matched funding from the government. We will stick to our guns and build the kind of prosperous community we have agreed that we want. We have nothing against business; we just want to do it our own way." This man clearly means business.

Later Impact on Reputation

Option A—Business as Usual

Media Reaction

There were several "farewell to BBJ" pieces in the local press and even some national coverage to this effect but, overall, the decision to scale back and depart has been a non-event in PR terms. Fujiwara Tatsuji's business column in the Mainichi Shimbun was an exception, in that he saw the move as an opportunity missed.

Option B—A Major Donation as A Final Gesture

Media Reaction

The fact that the Mayor's outstanding target of ¥72,000,000 was met within 24 hours of the call to arms by BBJ was impressive and got good coverage. It seems likely that BBJ will be fondly remembered for its role here as polling reveals a very positive public attitude toward the company.

Option C—Let the Locals Figure It Out

Media Reaction

Recent coverage of BBJ's work in Kyushu has focused on Mr. Reed's statements about the future. His indication that the company may well be interested in investment, depending on "the kind of new economy that the community wishes to build" has been widely quoted. The reaction has been mixed and, to a degree, puzzled. Green Action has led the charge, asserting that "companies such as BBJ will not be welcome in the new Kyushu eco-zone."

Option D—A Plastics Plant Delivered Through Pragmatic Compromises

Media Reaction

The press likes the idea of the Green Poly Plant; this positive coverage has helped to move local opinion, which had been skeptical.

Option E—A Development Council

Media Reaction

The publication of the North Kyushu Development Council's delivery plan, "Silver Lining," received strong coverage across the media. It was helpful to have BBJ alongside Green Action, Oxfam and others on a platform that was nonetheless chaired by a local fisherman. Warwick University's School of Business will shadow the Council during its first two years, in a move which lent considerable weight to the project and gave it a strong international dimension. The eyes of the world were certainly watching. CNN's hour-long special on the project and its genesis was itself an excellent stimulus for new investment.

General Impact on Firm Value[1]

Option A—Business as Usual

COMPANY REPUTATION: Brighter Brand Japan's decision to return to business as usual did not attract a great deal of attention at the time. A number of critics in the press saw it as an opportunity missed, but this view was hardly mainstream. Shares prices did not respond negatively. Meanwhile, the company still benefits in the afterglow of its philanthropic work, which remains fresh in the minds of the Japanese public. This resulted in a net value-added of $3,000,000.

SALES: Sales figures showed a healthy increase on the month, as BBJ was able to put its full weight behind production, marketing and distribution. It seems very clear that the company will shortly return to pre-quake levels of business, if not exceed them. One Board member wrote to the Chairman complaining that sales would be much higher were BBJ integrated into a rebuilding plan—but there seems little evidence for such an eccentric opinion. The net value added from returning to normal operations is $7,000,000.

RISK MANAGEMENT: A stormy meeting of BBJ's Risk Management Group concluded that the company's current posture was leading to increased risk exposure. The argument that prevailed was that the Japanese strategy was far too conservative. Rebuilding efforts in the Kyushu area would represent a not-insignificant piece of Japan's capital investment for next year, which BBJ was ignoring. The Group envisages a loss in market share in BBJ's construction arm, which was supposed to receive priority according to the current corporate plan. A push into Korea seems to be the alternative: but will require significant investment to build markets. The net effect is zero dollars of added value.

SAVINGS IN SUPPLY COSTS: BBJ now has its entire distribution fleet up and running again since its withdrawal from aiding the relief efforts in Kyushu and the company's return to business as usual. But commodity costs—especially for oil and plastics, remain sky-high. In addition, 30% of the fleet remains off the road at any one time, due to the backlog in routine maintenance. This has unfortunately brought close to zero for net value added in supply cost savings.

STAFF MOTIVATION: Motivation and morale appear normal in these early weeks after the earthquake. There is huge relief at the ability to return to normal life, but an edginess concerning the perception that this may also be an opportunity missed. Productivity, sickness and staff turnover are back to pre-quake levels. This has had a positive value-added gain of $4,500,000.

Option B—A Major Donation as A Final Gesture

COMPANY REPUTATION: Everyone was more than pleased with BBJ's offer to rebuild a school in the affected region of Kyushu. Our donation to the Mayor's Fund was matched by other MNCs and the remainder was raised within the day. Press coverage was positive across the board and the share price is up again. In reputational terms, this exit strategy has proved a winner. There is a positive $6,000,000 net value added.

SALES: Surprisingly, domestic sales have dipped in recent weeks. Our marketing department is blaming the major new advertising campaign by Easy Clean, coupled with the gloomy national economic outlook. This may be a blip, but it has added no value to our sales value for this period. We need to look at this.

RISK MANAGEMENT: Having by now priced in the new risks identified subsequent to the earthquake, BBJ's decision to cease its involvement in the area simplifies its risk assessment work. The company clearly faces a conventional commercial challenge in the form of Easy Clean, to which a marketing response has been developed, budgeted and reflected here. The possibility of lost investment opportunities in disengaging at this stage has also been costed. A modest net value of $3,750,000 has therefore been posted for risk.

SAVINGS IN SUPPLY COSTS: BBJ did not benefit from any savings in supply costs. In fact, costs are still quite high due to the damaged infrastructure and alternate routes required in order to get products where they need to go. Raw materials such as oil and plastics remain at a premium. All of BBJ's potential savings were eaten up. No net value added.

STAFF MOTIVATION: Rebuilding the school rewarded not only the company's public image, but also the employees of the company, many of whom want to continue helping with recovery efforts. They are also proud that BBJ was the catalyst to get the Mayor's Fund back on track. As BBJ swings back into business, as usual, the staffing legacy is impressive: a tight-knit, highly motivated workforce with a renewed faith in its employer. Metrics such as sickness and absenteeism levels all reflect this. The net value added for employee motivation is $2,500,000.

Option C—Let the Locals Figure It Out

COMPANY REPUTATION: When other MNCs went ahead with projects without gaining the support of the local community and NGOs, their images suffered. Brighter Brand Japan's decision to await a local consensus met with a positive reaction, even from some of our more outspoken naysayers. Importantly, this is being seen by the markets as a shrewd long-term strategy. Institutional investors have been buying into BBJ and pumping up the share value. Reputation is now valued at $8,000,000.

SALES: Sales continue to recover from the losses suffered during this crisis, as BBJ's delivery operation plugs the gap, and the goodwill secured amongst the public keeps the demand buoyant. The pause in activity vis-à-vis Kyushu has allowed an increased focus on this area, which for the short term has been beneficial. This has contributed to a positive net value of $5,500,000.

RISK MANAGEMENT: This is a conservative strategy, which has allowed us to scale back some of the costly risk mitigation activity which might have accompanied a bolder posture. Project insurance on a major construction scheme in the Kyushu area was estimated at $3,000,000, for example. The uncertainties of operating against this backdrop were just too great. Another was the prospect of a hostile environmental campaign from Green Action: something else BBJ no longer needs to counter. Consequently, we estimate a net value-added gain of $6,750,000.

SAVINGS IN SUPPLY COSTS: Material and transport costs in Japan remain at very high levels. Although BBJ has been able to restore pre-quake delivery and production schedules, it has done so at a high cost. The value added is zero.

STAFF MOTIVATION: Slowing the pace has enabled BBJ to pull people back from the immediate quake area, and we have seen a modest improvement in productivity and absence levels. A more ambitious scheme for working with the local community might have led to a more committed workforce, but this is largely abstract speculation on the part of the HR department. This positive uptick has led to $2,000,000 in value added.

Option D—A Plastics Plant Delivered Through Pragmatic Compromises

COMPANY REPUTATION: BBJ's reputation took a major turn upwards when it was announced that we had come to terms with Green Action to build an eco-friendly plastics plant on the Kyushu coast with environmentally safe transportation systems as well. Also boosting the reputation was BBJ's willingness to subsidize a health scheme for its workers. The positive effects on the company's reputation are valued at net $12,000,000.

SALES: Once the plant was up and running, its volume of sales was within the parameters the company projected. However, project delays —chiefly associated with the completion of the new railhead—impacted on this. The broader Japanese market remained buoyant. The estimated net value of sales for the period is a reasonable $6,000,000.

RISK MANAGEMENT: One of the most significant risks to BBJ's continued operations in the region was posed by Green Action. They threatened everything from sit-ins to boycotts. Eventually, both sides were able to meet in the middle with a plan that worked for both sides. A current risk is whether or not there will continue to be enough skilled laborers in the area to work at the new plant. BBJ's risk register has therefore had one major risk checked off during this period. Risk management techniques equated to $1,500,000 in value.

SAVINGS IN SUPPLY COSTS: The direct connection to distribution hubs has added to the savings in supply costs. However, BBJ had to wait six months after the completion of the factory before the railroad was up and running. This led to a decrease from the previous estimates and only achieved $2,250,000 in value added.

STAFF MOTIVATION: The Saiki project has demonstrated a promising new template for production within the BBJ family. The decision to involve trade unions and community leaders throughout the project lifetime was a shrewd one, in that there is now a genuine sense of ownership. Productivity and attendance at the new plant represent company records. Total added value is $4,000,000 due to staff motivation.

Option E—A Development Council

COMPANY REPUTATION: The Development Council established by all parties was a huge success and BBJ was glad to be on board. The likely (and unlikely) partnerships with NGOs, MNCs and the local government seem endless. This forward-thinking initiative increased the name recognition of BBJ right across the nation. This development model could serve well in other areas of Japan, where BBJ operates. Shares are up on the broad business opportunities that this project promises. The total value added is $9,000,000.

SALES: By being part of the North Kyushu Development Council, BBJ was able to build its advanced plastics plant at pace. This was in light of promised eco-friendly upgrades to be made to the current design, and it received the support of both the locals and Green Action. Consequently, BBJ has been able to launch new product lines in Asia faster than originally foreseen, with a resulting sales fillip. Production at the plant began earlier than planned and net value added to sales is $13,000,000.

RISK MANAGEMENT: Having factored in and mitigated most of the direct risks associated with the earthquake during the last quarter, the decision to join the Development Council merely embedded previous decisions. The risk value for this period is zero.

SAVINGS IN SUPPLY COSTS: Localized production of plastic packaging has already generated significant savings in supply costs, more so since the completion of the new rail link to Saiki. The value of this development is likely to increase, as BBJ ramps up production and is able to further reduce dependence on imported packaging. This equated to a value added of $3,500,000.

STAFF MOTIVATION: New staffing practices at the Saiki plant have developed such concepts as more flexible rostering in exchange for on-site childcare. Recruitment and retention have been exemplary, and for an area recovering from a disaster such as this, so have sickness levels. The intention is to learn from this experience and transfer some of these concepts to other parts of the business. The net value added for staff is deemed to be $2,500,000.

Reflecting on Chapter Three

The Scenario in Context

Set four months after the disaster in northern Kyushu, in this scenario teams were asked to consider strategies for BBJ as the authorities' thinking turned to future redevelopment. A key factor was the need for the company to build a new polythene factory. It had initially been assumed that this would be sited in Honshu. In light of the situation in Kyushu however, it made sense to switch locations. Alternatively, it might be more straightforward to regard the situation as "business as usual." In this case, there would likely be commercial opportunities arising from the redevelopment strategy. The idea of a large and final charitable gesture, such as the rebuilding of a school, had also been mooted. Thinking on this depended on a review of the degree to which BBJ intended to remain engaged with the Kyushu situation.

Dilemmas and Challenges

It was very evident from the Field Report and Pentalemma Briefing that there were strongly held views about a range of possible futures for the northern Kyushu area. As a result of this, there was perhaps a case for taking a "safety first" approach and steering well clear of controversy. Having contributed so much already, it would be regrettable if BBJ were now to attract a lot of negative publicity by wading into these arguments. The proposal for the polythene plant, for example, was opposed in principle by the powerful Green Action lobby group. They feared that such a development would entail significant levels of pollution in the area.

Was this necessarily the case? Also, was the stance taken by Green Action as absolutist as it might have seemed? Although it is sometimes valid to walk away from percussive situations such as this, it is often the case that careful preparatory work will point to promising ways forward. An emphasis on identifying stakeholders, vested interests and different positions is a key aspect of such preparation. The interests and ambitions of the other commercial organizations involved could have been a fruitful area of research. The likelihood of supplier or customer industries co-locating in a rebuilt Kyushu enterprise zone and similar intelligence might have tipped the balance one way or another. Other inquiries could have been made about critical factors such as the labor market, proposed new transport links and so forth.

A corollary to this point concerns the conduct of any resulting negotiations. Companies finding themselves working with tough-minded stakeholders in crisis situations can take measures to make the going smoother for all.

BBJ faced a potentially tense series of discussions surrounding the future of northern Kyushu, which it might have legitimately avoided. However, approached in the right spirit and using the right techniques, such situations can generate creative solutions of benefit to a wide range of stakeholders.

Note

1. The scenarios in this book are hypothetical, so results are based on plausibility. In general, there are three categories of priorities that I focused on by using John Elkington's concept of the triple bottom line; the idea of balancing financial, social and environmental priorities (Elkington, 1999). As one soon discovers, however, any attempt to balance these inevitably leads to dilemmas. Much of the creativity needed to handle these dilemmas was inspired by Johan Galtung's model of conflict transformation (Galtung, 1996).I studied a wide range of companies that have faced dilemmas in pursuing these priorities, drawing especially on Jane Nelson's research of businesses in high-risk environments (Nelson, 2000). Adam Kahane's Mont Fleur scenarios informed the storyline by providing a firsthand review of the planning process inside companies operating in conflict areas (Kahane, 2012). On the whole, extrapolating from these cases, the results for the majority of the stakeholders are usually unsustainable when their decisions only serve a minority of their stakeholders. The results are also negative when decisions are delayed or not implemented. The results tend to offer more lasting value when decisions satisfy the needs of most of the main stakeholders reasonably well over longer timeframes.Finally, to apply the lessons learned from Elkington, Galtung and Nelson to practical management challenges, Ed Freeman's work on Managing for Stakeholders provided helpful examples of how everyday management dilemmas can be reframed as resource issues between stakeholders (Freeman, 2007). The scenarios have drawn on lessons learned from these and other scholars and practitioners to create the heuristic process used for developing plausible outcomes to the various dilemmas presented in this book.

Further Reading

Elkington, J. (2002). *Cannibals with forks: the triple bottom line of 21st century business.* Oxford: Capstone.

Freeman, R. E., Harrison, J. S., & Wicks, A. C. (2007). *Managing for stakeholders: survival, reputation, and success*. New Haven, CT: Yale Univ. Press.

Galtung, J. (2012). *Peace by peaceful means peace and conflict, development and civilization.* Oslo: Prio.

Kahane, A. (2012). *Transformative scenario planning working together to change the future.* San Francisco, CA: Berrett-Koehler.

Klein, N. (2007). *The shock doctrine*. London: Penguin Books Ltd.

Nelson, J. (2000). *The business of peace: the private sector as a partner in conflict prevention and resolution*. London: International Alert.

Žižek S. (2018). *First as tragedy, then as farce*. London: Verso.

Conclusion

Highlights from Chapter One

In Chapter One, certain options were made available to BBJ managers during the intense period of their initial disaster response. The scenario suggested that Kelvin Reed had taken matters into his own hands by authorizing BBJ's lorries to hit the road immediately after the earthquake. Switching a critical component of a company's infrastructure to emergency relief work in this way would almost certainly not be within the rule book. Corporate governance cannot legislate for every eventuality, and senior managers are paid to manage. In the real world, there needs to be sufficient flex within such arrangements to allow a company to respond to a genuine emergency. This was very much the case in Japan in 2011—to the credit of the companies concerned. What remains important under such circumstances is that the manager concerned is able to justify his actions and that he seeks retrospective authority as soon as practically possible. Due process is very important; but in crisis situations, managers must be prepared to set it aside where necessary—and be ready to defend their decisions.

Regarding philanthropy, the temptation to view this situation as a straightforward moral case for philanthropic business support is obvious and entirely defensible. What the scenario endeavors to show though, is that there are often creative solutions that go further, benefitting both the community and the business and perhaps with an eye on the longer term. The "Full Philanthropic Engagement" choice, although laudable, does not consider these broader questions. Whilst BBJ's reputation will soar, sales will drop due to logistical problems, and supply costs will go up. It is far from certain that this is the optimum choice here. Other options—notably "Smarter Relief" and "Partnership Based Commitment" also included a strong element of continued engagement by BBJ. By contrast, "Scaling Back" would have equated to withdrawal, with some costly implications for share price and reputation. A rejection of philanthropy, therefore, would have been a mistake. But an open-ended commitment with no thought for the future would have been to miss significant opportunities.

What set aside the "Partnership Based Commitment" option from the others was its creative use of the other players involved in the emergency. This served to maximize impact and minimize cost. It also put a marker down for any future engagement. The consequence for BBJ was protected sales, reduced risk costs as risks are scoped and shared, and a stellar reputation. By far the best option in this scenario, this choice showed that it is possible for companies to respond charitably to a civic emergency and in so doing control their exposure and perhaps create future opportunities: A textbook win-win situation.

Highlights from Chapter Two

In Chapter Two, some of the more important themes that this scenario was designed to highlight are considered below, in the context of the available options. It is hoped that this will serve both for future reference and also as a further explanation of the linkage between the dilemmas and outcomes enshrined in Pentalemma.

One of the key concepts here was to build upon the early success of partnerships with other companies involved in the disaster relief operation. This was the way to protect sales and yet continue to ensure that BBJ secured a major impact on the ground in northern Kyushu. The "Full Integration" strategy was the best available precisely because it exploited this approach to the full. KZ Foods would do the heavy lifting, freeing up BBJ's lorries, with the result that sales recovered. The creative delivery of services on the ground through partnership also generated much better risk analysis (and therefore control), as well as an improved reputation and great press coverage, whereas a "Return to Commercial Priorities" would have effectively abandoned such partnership work. Reputation, sales and staff morale would have plummeted on the back of a superficially attractive strategy, based on firefighting the immediate sales problem in Kyushu.

None of the above analysis is supposed to imply that managers should switch off their commercial radar during a crisis of this kind. The offer made by CS was a genuine threat. It did need to be looked at in commercial terms, alongside the other factors in this difficult situation. This is not inappropriate. BBJ obviously enjoyed a degree of goodwill and a solid reputation in this area of civil engineering. All of these circumstances should have signaled the advantages of remaining engaged. "A Measured Approach" was one of the better strategies available here, in that it acknowledged this commercial dimension and protected sales. The absence of a partnership dimension was perhaps its drawback, meaning that the supply problem would have remained critical. There is actually no inherent contradiction between engaging positively in partnership with what might be commercial rivals and recognizing that in other circumstances, their activities pose a risk to one's own business which needs to be assessed. It simply depends on the different options available. The fact of a crisis situation does not suspend the normal rules of commercial competition.

Regarding public-private partnerships, the models for engaging with the public sector in capital projects are increasingly varied. The one outlined in the "Full Integration" option is only illustrative, as well as being drawn up in the particular context of a crisis situation. The point remains valid and is, in fact, broader: where public and private interests converge, government agencies will often be open to creative funding models reflecting this. They might split resource and capital costs (as here), entail different ownership models such as long-term leasing or lease-back, or any number of other possibilities. Creativity and open-mindedness, coupled with a recognition of both parties' constraints and ambitions, is crucial.

Highlights from Chapter Three

In Chapter Three, some of the more important themes that this scenario was designed to highlight are considered below, in the context of the options available to managers. This serves as a future reference and as a further explanation of the linkage between the dilemmas and outcomes enshrined in Pentalemma.

The main opportunity that this scenario presented to BBJ was that of ensuring its involvement in broad planning decisions for the northern Kyushu area. Put differently, if BBJ were to opt for a plastics plant in the disaster area, then it needed to maximize the benefit of this by influencing the macro plan for the region. In this way, it would be most likely to secure transportation, supplier industry, workforce and other advantages. It needed to be involved in the whole plan. This is why, although not a bad option, the "Pragmatic Compromises" choice generated only limited benefits in areas such as reputation and staff morale. The big payoff came with the "Development Council"—placing BBJ at the heart of the planning process. Large scale location decisions of this type require breadth of thought. The obvious economics associated with factors such as current transportation arrangements and market location may look very different if there is an opportunity to join an ambitious redevelopment plan.

In connection with the role of business in society, working with a community redevelopment program is not a one-way street. While its potential for commercial reward is high, business would be wise to regard it as a mutually beneficial partnership, rather than merely a vehicle for profit. The Development Council option is clear on this point. This is a more sophisticated approach to social engagement, with an evolving relationship between all stakeholders. Businesses cannot help but be involved in the societies within which they operate. Kelvin

Reed's point about this is a truism. In this scenario, the "cut and run" mentality implied by the "Business as Usual" option led to a marked reduction in business for BBJ's construction arm, as the firm lost out on big contracts in the Kyushu area. Businesses and society are integral to one another and should work toward mutual support. Not to do so invites ill will—and can quickly impact the bottom line.

Many businesses remain nervous about the "Green Agenda" and view it as a brake on profitability. Yet these are issues that citizens care about deeply. They cannot be ignored. On a more pragmatic note, common ground is often to be found. In this scenario, it proved possible to design a "clean" polythene plant which would generate local employment, have a neutral impact on the environment and solve BBJ's location dilemma. The seemingly stubborn Green Action group proved open to such creative thinking and took a similar stance with the thorny question of the local fishing industry. Activists and campaigners in this sector should be treated as legitimate stakeholders. In the end, compared to the other approaches available, making an effort to search for a positive-sum outcome creates more value. How this greater value is then distributed is a discussion for another day.

Other Books by This Author

Conflict Minerals in the Democratic Republic of Congo

In the Democratic Republic of Congo, you advise Speedy Devices, one of the world's largest consumer electronics producers. You will acquire a range of risk management tools as you help your client resolve complex issues related to profit, child labor and the role of business in a conflict situation.

Mining Conflict in the Philippines

Welcome to Mindanao, an island to the south of the Philippines. You will help Massive Metals, a foreign mining company, to manage its stakeholders. You will learn to use problem-solving tools to resolve environmental and political issues.

Coffee and Conflict in Colombia

You and your team advise Caffeine Supply, a global coffee exporter. The company has a successful history in the sector, but the industry is now facing a decline at home and competition from abroad. You will learn about fair trade and the basics of negotiation in a cross-cultural context.

Notes to Self

Chapter One

1. Who are the main stakeholders?

2. What are the stakeholders' main interests?

3. Which issues do the stakeholders disagree about?

4. Which of the above issues is the most critical one?

5. If the most critical issue is not resolved, what might happen to the stakeholders and your client?

6. Which option do you think is the best one for your client to choose?

7. How will the different stakeholders benefit from choosing the option you suggest?

Chapter Two

1. Who are the main stakeholders?

2. What are the stakeholders' main interests?

3. Which issues do the stakeholders disagree about?

4. Which of the above issues is the most critical one?

5. If the most critical issue is not resolved, what might happen to the stakeholders and your client?

6. Which option do you think is the best one for your client to choose?

7. How will the different stakeholders benefit from choosing the option you suggest?

Chapter Three

1. Who are the main stakeholders?

2. What are the stakeholders' main interests?

3. Which issues do the stakeholders disagree about?

4. Which of the above issues is the most critical one?

5. If the most critical issue is not resolved, what might happen to the stakeholders and your client?

6. Which option do you think is the best one for your client to choose?

7. How will the different stakeholders benefit from choosing the option you suggest?

www.ingramcontent.com/pod-product-compliance
Ingram Content Group UK Ltd.
Pitfield, Milton Keynes, MK11 3LW, UK
UKHW021828100926
8556IPUK00003B/213

9 781433 175305